The Herds Shot Round the World

FLOWS, MIGRATIONS, AND EXCHANGES

Mart A. Stewart and Harriet Ritvo, editors

The Flows, Migrations, and Exchanges series publishes new works of environmental history that explore the cross-border movements of organisms and materials that have shaped the modern world, as well as the varied human attempts to understand, regulate, and manage these movements.

The Herds Shot Round the World
Native Breeds and the British Empire,
1800–1900

Rebecca J. H. Woods

The University of North Carolina Press CHAPEL HILL

The publication of this book was supported in part by a generous gift from Cyndy and John O'Hara.

© 2017 The University of North Carolina Press
All rights reserved
Set in Espinosa Nova by Westchester Publishing Services

The University of North Carolina Press has been a member of the Green Press Initiative since 2003.

Library of Congress Cataloging-in-Publication Data
Names: Woods, Rebecca J. H., author.
Title: The herds shot round the world : native breeds and the British empire, 1800–1900 / Rebecca J. H. Woods.
Other titles: Flows, migrations, and exchanges.
Description: Chapel Hill : University of North Carolina Press, [2017] | Series: Flows, migrations, and exchanges | Includes bibliographical references and index.
Identifiers: LCCN 2017019373 | ISBN 9781469634654 (cloth : alk. paper) | ISBN 9781469634661 (pbk : alk. paper) | ISBN 9781469634678 (ebook)
Subjects: LCSH: Animal industry—Great Britain—History—19th century. | Sheep breeds—Great Britain. | Cattle breeds—Great Britain. | Endemic animals—Great Britain. | Great Britain—Colonies—History—19th century. | Industrial revolution—Great Britain—History—19th century. | Agrobiodiversity.
Classification: LCC HD9436.G72 W66 2017 | DDC 338.1/76082094109034—dc23
LC record available at https://lccn.loc.gov/2017019373

Cover illustration: Large flock of merino sheep, South Australia, ca. 1900. National Library of Australia, nla.obj-144081153.

Portions of chapter 4 were previously published in a different form as "From Colonial Animal to Imperial Edible: Building an Empire of Sheep in New Zealand, c. 1880–1900," *Comparative Studies of South Asia, Africa, and the Middle East* 35, no.1 (2015): 117–36; and "Breed, Culture, and Economy: The New Zealand Frozen Meat Trade, 1880–1914," *Agricultural History Review* 60 (2012): 288–308. Used here with permission.

For my parents

Contents

Acknowledgments xi

Introduction 1

PART I | *Great Britain*

    1   A Breed in Any Other Place 25

    2   Much Ado about Mutton 52

    3   The First Breed of Cattle 78

PART II | *Greater Britain*

    4   Native Colonials 109

    5   A Universal Type 140

        Conclusion 165
        The Return of the Native Breed

        Notes 177
        Bibliography 213
        Index 227

Illustrations

Map of the British Isles 26

Sheep breeds characteristic of the "Higher Welsh Mountains" 32

The New Leicester or Dishley breed of sheep 42

Black-faced heath sheep 44

Southdown sheep 50

Merino ram 53

A pair of merino sheep 57

A pair of Exmoor sheep 68

The Ryeland breed of Herefordshire 75

A Hereford bull 80

A Hereford cow descended from the "Tomkins blood" 84

A Shorthorn bull 91

A Longhorn bull 92

A bull of the "wild" white type 94

Corriedale sheep 110

Display of frozen sheep carcasses outside the British New Zealand Meat Company, Christchurch, c. 1900–1920 127

Large flock of merino sheep, South Australia, c. 1900 128

Corriedale-cross sheep at Cheviot Hills Station, Canterbury, c. 1893 137

A Shorthorn cow 142

Portrait of Sir Bartle Frere 158

A Traditional Hereford cow 169

The entrance to the Hereford Cattle Society offices 171

GRAPHS

British wool imports in millions of pounds, 1790–1870 61

Tons of frozen sheep meat imported to Great Britain, 1885–1910 121

Acknowledgments

The sources of aid and assistance, both material and intellectual, that I have benefited from over the course of this work's development from the germ of an idea to published monograph are many. First and foremost among the debts I have incurred is that which I owe to Harriet Ritvo, without whom this work would scarcely have been conceived, much less completed. She has been an unparalleled mentor, advisor, and friend throughout. Without Alistair Sponsel, this book would not be so aptly titled. Early on, the support of MIT's History, Anthropology, and Science, Technology and Society (HASTS) graduate program and of several funding agencies made it possible for me to undertake the work necessary to complete its first iteration as a dissertation. Preliminary research in the United Kingdom was funded by a Dissertation Proposal Development Fellowship from the Social Sciences Research Council (SSRC), while an extended research trip to New Zealand and the United Kingdom in 2009–10 was possible thanks to an International Dissertation Research Fellowship (SSRC) and a Fulbright Fellowship to New Zealand (Fulbright-IEE and Fulbright New Zealand). Completing the final stages of the project would hardly have been possible without the benefit of a Mellon-ACLS Dissertation Completion Fellowship awarded by the American Council of Learned Societies in 2012. The History Project at Harvard University and the University of Cambridge also generously funded a final research trip to the United Kingdom in April 2013.

Practical and material support from these funding agencies made it possible for me to undertake the work necessary for executing this project; the intellectual and moral support of mentors and colleagues in the HASTS graduate program at MIT were no less integral to the early phases of this project. David Jones and Janet Browne at Harvard University were essential to the formulation and execution of this project at its earliest stages, and I am grateful for their support. Chris Capozzola, Stefan Helmreich, David Kaiser, Leo Marx, and Heather Paxson each made my time at HASTS intellectually richer. Etienne Benson, Lisa Messeri, Nadya Peek, Sophia Roosth, Shira Shmu'ely, David Singerman, Alma Steingart, Michaela Thompson, and Emily Wanderer have been valued interlocutors over the years.

I also owe a special thanks to Bill Turkel at the University of Western Ontario, who set me on the path to graduate school, and who gave me necessary early encouragement. Parts of this work benefited from feedback received at several conferences and workshops, including the HASTS Program Seminar (2012), the Dissertation Proposal Development Fellowship Animals Studies group (June and September 2008), the International Workshop for PhD Scholars in Environmental History (Australian National University, October 2010), the International Dissertation Research Fellowship workshop (March 2011), the Harvard Symposium on Energy and Environment in Global History (Harvard University, April 2011), the German Historical Institute Conference on Global Cash Crops (June 2011), and the Nonhuman Empires: Between Agents and Actants Conference at the Department of History and Philosophy of Science (University of Cambridge, March 2012). I especially thank Jim Secord and Thomas Dunlap for their intellectual contributions to my work.

The Columbia University Society of Fellows in the Humanities provided an ideal intellectual home in which to transform this project into a book. The intellectual community I found there has made this project a richer and more mature one than it otherwise would have been. Thanks to my fellow fellows Vanessa Agard-Jones, Teresa Bejan, Ben Breen, Maggie Cao, William Deringer, Brian Goldstone, David Gutkin, Hidetaka Hirota, Murad Idris, Ian McReady-Flora, Dan-el Pedilla Peralta, Carmel Raz, and Grant Wythoff. Thanks also to Karl Apphun, Christopher L. Brown, Deborah R. Coen, Isabel Gabel, Matthew Jones, Eugenia Lean, Malgorzata Mazurek, Susan Pedersen, Pamela Smith, Deborah Valenze, and Carl Wennerlind. And a special thank-you to Eileen Gilooly for her unstinting support and for keeping the whole show running.

While researching this book, I benefited greatly from the assistance and expertise of a number of librarians, archivists, and livestock breeders close to home and throughout the United Kingdom and New Zealand. Critical support came from the librarians and archivists at the British Library; the Walter Frank Perkins Agricultural Library (Southampton University); the Museum of English Rural Life (University of Reading); the University Library (University of Cambridge); the National Archives of Scotland; the Herefordshire Record Office; the Lincolnshire Record Office; the Derbyshire Record Office; the Hereford Cattle Society; the Lincoln Sheep Breeders' Association; the Rare Breeds Survival Trust; the library of the Royal Agricultural Society of England; the Australian National Library; Archives New Zealand and the Alexander Turnbull Library in Wellington,

New Zealand; and the Hocken Library (University of Otago). Special thanks to John Charles, Associate University Librarian at Massey University (Palmerston North, NZ) and to Michelle Baildon at MIT. Les Cook, Peter Talbot, and Bev Trowbridge were among the practical breeders who generously shared their time and knowledge with me.

The finishing touches to this manuscript took place at the University of Toronto's Institute for the History and Philosophy of Science and Technology. I thank the Institute's HPS community—especially Joseph Berkowitz, Lucia Dacome, Yiftach Fehige, Craig Fraser, Muna Salloum, Mark Solovey, Wen-Ching Sun, Marga Vicedo, and Chen-Pang Yeang—for their warm welcome.

Thanks to everyone at the University of North Carolina Press for the execution of this publication, especially the series editors Harriet Ritvo and Mart Stewart; and Brandon Proia and his team.

Finally, I thank my family—my brothers, Ned and Alex; and my parents, Betty and Frank—for their love, support, and patience. Since beginning this long intellectual journey, I have been fortunate enough to find Gabe Friedman, a now constant companion whose presence, love, and generosity has transformed all aspects of my life. Thank you, Gabe.

The Herds Shot Round the World

Introduction

At the outer edge of the North Atlantic continental shelf, fifty miles of sea between it and the Outer Hebrides, lies St Kilda, a small archipelago of volcanic origin that is home to an unusual ovine population alleged by conservationists and scientists to be "the most primitive domestic form [of sheep] in Europe."[1] Soay sheep, so called after their original islet, are dark brown or buff in color, horned, and celebrated for their resemblance to *Ovis musimon*, or the mouflon, the nearest wild relative of *Ovis aries*. Archeological evidence seems to support this contention, as do the color-markings of a subset of the population, whose white bellies and dark brown coats—an unusual combination among domesticated ovines—make them strongly reminiscent of the mouflon's own appearance. Add to this the fact that Soays are self-shedding (a common feature of wild ovines) and "behave much more like wild animals than modern domestic breeds,"[2] and it is easy to see why ecologists, environmentalists, and members of the public believe Soays are "the only *living* remnant" of "those first civilized cultures of our islands," surviving the passage of time relatively unchanged since Neolithic farmers and fishers are believed to have (temporarily) settled the remote islands.[3]

Soay sheep were also among the first domesticated animals targeted for conservation in the twentieth century: in 1932 the laird of St Kilda relocated 107 sheep from their eponymous home to neighboring Hirta in an effort to establish a satellite population as insurance against the possibility of the breed's extinction on Soay.[4] At the same time, a handful of wealthy mainland conservation-minded eccentrics began taking up "some of the wild sheep of Soay" as park animals, establishing them on great estates dotted throughout England and Wales, again, "in case they die out on Soay."[5] By the 1970s, relocating portions of populations of numerically challenged, unusual, or historically interesting types of livestock in this way as insurance against future threats had become standard practice within a growing popular movement to preserve and protect rare or endangered breeds of livestock in Great Britain. Britain's first and foremost organization for breed conservation, the Rare Breeds Survival Trust (RBST), worked zealously in the 1960s and 1970s to relocate populations of insular or otherwise isolated types of sheep to and from various corners of the United Kingdom.[6]

Evidence supporting the efficacy of such action accrued: the number of Soay sheep grew, for example, from an estimate of about 500 confined to St Kilda in 1939 to several thousand scattered throughout the United Kingdom in 2016.[7]

But already by 1974, more reflective members of the newly formed organization were beginning to wonder whether in fact such a policy posed an inherent problem for the RBST. J. C. Hindson, a founding member of the organization and a veterinarian, worried that the trust's practices "involve[d] the essence of its own destruction" because of the "inevitable change in any breed or species which will occur when a change of environment takes place."[8] Given the right set of circumstances, enough time, or a lapse in conservationist vigilance, this "inevitable change" might, he feared, alter a breed beyond recognition. Such a possibility, by extension, called into question the organization's own rationale—that of preserving rare genetic traits for possible future utility, or for their historic significance. More than this, it called into question the notion of a breed itself: to what degree was a type dependent on its environment for its defining characteristics? And did the inevitable tendency of a type to change over time in response to its surroundings mean that the very existence of breeds was a fiction? Or, "posed as a simple question," (as Hindson asked) "is a Soay still a Soay after 25 generations in the South of England?"[9]

In fact, this question is anything but simple. It conjures larger issues about the roles for heredity and environment in shaping a breed, and the degree to which humans may intercede in, and modify, the descent of other kinds—concerns well known to British breeders since at least the eighteenth century, and particularly pressing in the context of nineteenth-century imperial expansion. Hindson's question, moreover, raised the specter of indigeneity, the criteria by which certain kinds are (or are not) deemed "native" to a place, and the fragility of this category itself. After all, if the British Isles' most aboriginal type could be irrevocably altered by simple regional translocation, how durable could nativeness be, and how much more perilous cultural and environmental transpositions on the imperial scale?

These are the concerns that guide this book. It takes as its subject the problem of the Soay sheep writ large. Where Soays moved from farthest Scotland to the South of England, British breeds moved from the United Kingdom to the farthest reaches of the "New World"—North and South America, Australia, New Zealand, South Africa.[10] Where Soays moved from a harsh and challenging environment to gentler, more luxurious surroundings, imperial breeds left the soft cradle of their home islands for an enormous range of conditions, some of them equal to or surpassing their native pastures in comfort, many of them not. Where breeder-conservationists feared the loss of Soays' unique, island-bred characteristics, colonial breeders feared the loss of Britishness in their stock, and by extension, even themselves. The persistent difficulty of sorting out the influence of artificial selection at the hands of their human keepers from the "natural" effects of environmental pressures ensured that the puzzle of whether or not British sheep in colonial Australasia, or British cattle in the nineteenth-century Americas, remained British, continued unresolved. The question of whether a Soay sheep, transposed to the South of England, was still a Soay, remained unanswered.

THIS IS A story about the rise and fall of British breeds, about the invention and reinvention of "native" types of sheep and cattle, about the roles they played in the economics and ecologies of empire in the nineteenth century, and about their signification in postimperial British agriculture. From the midlands of England to Southland, New Zealand; from the county of Herefordshire to colonial Australasia, North America, and back; this book follows the fates of native British breeds across time and space. It begins with their formulation at a moment of agricultural rationalization, at the turn of the nineteenth century, when the regional variety of the British Isles' domesticated types became the raw material out of which agriculturalists crafted "improved" types in support of an industrializing society's growing alimentary needs. These improved breeds were subsequently enlisted as hoof-soldiers in the great agropastoral expansion of the British Empire from the mid-nineteenth century to its close, contributing to both the rationale and the material presence necessary for the expansion of British interests, capital, and settlers into previously un- or otherwise-occupied territories, all in the service of feeding the great maw that was industrial London. Later, British breeds fell from prominence in the changing conditions of post–World War II global agricultural production, losing ground globally and even at home to superproductive European and erstwhile colonial types, before they finally returned to public attention within the context of rare breeds conservation efforts in the late twentieth century.

To tell this story, this book queries the role, importance, consequences, and shifting meanings of "native" British breeds of sheep and cattle in the colonial project of the nineteenth century. At its core are questions about the relationship between type and place: what role do cultural and economic factors play in establishing the discursive connections between breed and environment? What happens when transplanting stock from one place to another, be it within the British Isles or across hemispheres, severs the links thus constructed? What does it mean to call a domesticated type of animal "native" to a particular place, and in the context of particular (political, economic, and cultural) discourses? Answers to these questions bear not only on issues of production and commerce, ecology and conservation, theories and understandings of heredity. They bear on the nature of empire itself.

The Rise and Fall of British Breeds

By the close of the nineteenth century, British breeds of livestock had conquered the world. Lincolnshire sheep grazed happily on the Canterbury plains of New Zealand, while Hereford bulls throve as well on the rough scrubland of Queensland, Australia, and the snow-covered pastures of Western Canada, as they did in the green paddocks of their native Herefordshire. These breeds dominated a system of pastoral economies in which Britain's appetite for meat and capitalism's appetite for profit drew people, animals, and terrains into a tightening web of production. Their ubiquity was part of a mode of livestock production that exceeded the contours of the British Empire, engaging colonial and quasi-colonial places alike in the nineteenth century, and establishing patterns of industry that laid the foundations for modern globalized meat production.

As the British Empire's fortune waxed, so too did those of its flagship breeds—the numbers of Hereford, Shorthorn, and Angus cattle, not to mention Romney, Ryeland, and Cheviot sheep, expanded throughout what is now the developed world. Yet by the late twentieth century, the position of such breeds had altered dramatically. Changing global patterns of production had motivated new developments in breeding. High yields and high productivity in particular became the watchwords of postwar farming in the twentieth century, and many of the British types, developed for a different set of market imperatives in the nineteenth century, fell from favor. Once mighty symbols of British influence around the world, Lincoln sheep, Shorthorn cattle, and others of their brethren were now rare even within the British Isles, treated with anxious care and the kind of nostalgia that echoed discourse surrounding endangered wild species.

The story of this rise and fall begins in the early nineteenth century. Part 1 of this book, *Great Britain*, locates the germ of a global system of imperial production in the dislocation of regional and subregional breeds of livestock within the British Isles. As Britain industrialized, and with the enthusiasm for agricultural improvement inherited from the previous century, breeders refashioned their livestock in the service of modern production. They worked to convert the regional or even local specificity of their animals into a fungibility that would enable stock to circulate more widely throughout Britain, answering the demands of an increasingly stratified system of production—in which the breeding of animals was segmented from their rearing and "finishing"—for standardized, geographically transposable types. And yet breeders could not afford to compromise the distinctive character of their breeds. Thus they walked a fine line between improvement and faithfulness to their breeds' identities, seemingly incompatible aims that nonetheless guaranteed the economic viability of their stock. National, and indeed nationalistic, appetites fueled this process, as a growing population in Britain demanded to be fed with ever increasing quantities of mutton, lamb, and beef. Consequently, throughout the United Kingdom, breeders reoriented local ovine and bovine variability toward market standardization, transforming cattle and sheep adapted to regional environments into bulky, quick-fattening beasts. The resultant breeds remained recognizably of a type—exhibiting very similar phenotypic features, behavioral patterns, or both—and yet they were geographically transposable, suited for increasingly sophisticated economies and industrialized production.

In the course of these developments, questions arose over type, place, and the proper method of selective breeding. Holding both the aims of improvement and breed identity in the balance challenged conventional wisdom, which held that "every soil has its stock,"[11] and the degree to which breed depended on environment, as well as how much humans could or should attempt to reshape local or regional varieties of stock, was increasingly a matter of debate. Conflicts arose over the relation of type to place; the role of the environment in shaping stock; the degree to which such effects were heritable; and how best to manage and manipulate the genetic potential of their stock. Theory and practice were debated—sometimes politely and sometimes hotly—in the published treatises of specialists in selective breeding and in the agricultural press. Practical breeders, gentlemen farmers, and agricultural worthies debated optimal levels of consanguinity within a breed, the degree to which kinds could or should be mingled, and the means by which heritable traits, or the "character" of a breed, could best be molded. For the most part, these questions remained unresolved, signaling the degree to which the relationships between type, place, and heredity were constantly shifting in the nineteenth century.

With Great Britain's imperial expansion in the middle decades of the nineteenth century, the trends set in Britain in the early part of the century played out on a grander scale. Part 2, *Greater Britain*, follows the herds and flocks of the British Isles out into the empire, tracing similar tensions between standardization and specificity in the livestock and landscapes of colonial Australia and New Zealand, and other places within Britain's sphere of influence in the Americas. The same tension between standardization and specialization that marked stockbreeding in Britain in the first half of the century shaped the dispersal of breeds throughout the empire, but questions about type and place became vastly more complicated in novel colonial settings during the latter half of the century. Here, stockbreeders served two masters: the unfamiliar climates and topographies of Australia, New Zealand, and North America, which demanded local adaptations; and the British consumer, whose dinner table was the end of the line for the bulk of colonial beef and mutton. As they tried to balance local adaptation and metropolitan taste, breeders experimented with heredity, testing the limits of contemporary understandings of heritability and breed plasticity. They developed new strains of livestock—such as New Zealand's Corriedale sheep, which was genetically derived from British breeds but culturally, economically, and environmentally hybrid—designed to grow British meat in the absence of British pasture.

Importantly, the development of pastoral economies in colonial Australasia and the Americas in the later nineteenth century depended on novel technological developments that lent the region unprecedented geographical and temporal proximity to the metropole. Foremost among these was mechanical refrigeration, which made the transport of dead meat from these places newly profitable in the late 1870s and early 1880s. The ability to store flesh almost indefinitely in a frozen state refigured colonial pastoral economies, and with them the breeds of empire. Sheep-producing areas turned from an exclusive focus on wool to growing meat as well. In cattle country, ranchers began an intensive process of "grading up" their motley herds with the use of imported British bulls, changing the bovine demography of the New World whether these animals trod the grasslands of Great Britain's overseas dominions proper or not. In these ways, bodies of sheep and cattle were remade to suit the refrigerated holds of ocean liners, and the empire itself was recast (at least in part) as a vast apparatus for feeding Britons.

The global expansion and dominance of British breeds was contingent upon a particular set of historical circumstances—the technological, economic, and environmental conditions that shaped imperial agroexpansionism in the second half of the nineteenth century. Without these conditions, the fate of purebred British domesticates was very different. Observers could still boast about the superiority of British breeds and the superlative skill it took to hone them as late as the mid-twentieth century. In 1949, for instance, the Earl of Halifax's foreword to a promotional handbook touting national types of livestock praised the "unsurpassed" skill of British breeders when it came to "the production of animals of the highest class."[12] But Britain's reign as the "stud stock capital of the world" was coming to an end.[13] In the postwar period, changing global patterns of production had motivated new developments in breeding away from long-standing British aims. Changes to British agricultural production and policy reflecting generalized anxiety over food security in the aftermath of World War II induced a preference for foreign or "continental" breeds: Dutch Holsteins increasingly took the place of Shorthorns in the milking stall; and beef breeds like Limousins, Charolais, and Belgian Blues replaced Devons and Lincoln Reds in the paddock.[14]

Under the accelerated pace of postwar production, the ability of continental breeds to put on flesh quickly was an asset, not only to pedigree breeders but to the whole chain of meat production in Britain, from finishing to butchering to marketing. At the same time, consumer tastes were changing, and the low butterfat content of Holstein milk, for instance, as compared with that of Alderney or Guernsey cows, likewise came to be seen as an advantage to the imported European breed. Moreover, livestock production in Britain at this time was becoming more intensive, and the abundance of fertilizers, feeds, and supplements meant that many of the characteristics that had made a "traditional" breed like the English Hereford or the Lincoln Red desirable—hardiness, independence, ability to forage, and the like—were no longer as relevant as they had been. As "continental" varieties moved into British pastures, they also increasingly took the place of British types in much of their erstwhile overseas territory.

Even where "British" breeds continued to dominate as, for example, the Hereford breed did in the North and South American cattle industries, they did so largely outside the grasp and influence of British breeders. The massive outflux of British bloodstock in the late nineteenth century had profited British breeders in the short run, but in the long run it enabled breeders in other places—New Zealand, Canada, the United States, even Argentina—to establish themselves as entrepôts for bloodstock in their own rights. What had been the strength of British breeds in the nineteenth century—their wonderful versatility, enabling them to thrive "both in Eastern and Western Hemispheres . . . in new homes and in surroundings very different from those in which they were bred and reared,"[15] as Halifax had put it—became another risk to British types at home. Even those that seemed secure like Hereford cattle, whose numbers largely withstood the foreign influx, were soon perceived to be at risk from reimported formerly colonial bloodlines that threatened to dilute the purity of "English" bloodlines. Already by the 1960s, people in Britain—some of them livestock breeders themselves, but many of them simply concerned or interested supporters of the countryside—became alarmed at a general loss of breed diversity within the nation's herds and flocks. Former champions of British stockbreeding took their place alongside Soay sheep as endangered types needing careful protection and preservation, as the nation wrestled with the dissolution of its empire more broadly.

The Breeds of Empire

The role of nonhuman creatures in the imperial triumph of the West, from four-footed domesticates all the way down to microbial carpetbaggers, has attracted much scholarly attention as a particularly effective way to counter claims of European technological and cultural superiority.[16] The theory of ecological imperialism holds, among other things, that differential disease environments were as significant as military tactics, mosquitoes as important as muskets, in determining the outcome of European contact with the Americas since 1500. In classic accounts like Alfred Crosby's *Ecological Imperialism* (1986), domesticated livestock played no small role in this process. For instance, in Eurasia, the tendency for people to live in close proximity to their cattle, pigs, and chickens contributed to the evolution of virulent crowd diseases. In turn, colonizers' acquired immunity to these diseases worked in their own favor in the New World, selectively sparing them as Old World epidemics devastated populations of native peoples previously unexposed to measles, influenza, smallpox, and other common (by European standards) diseases.[17] No less consequentially, Europe's long history of domestication also produced hordes of four-footed colonizers, whose grazing habits significantly altered the ecologies of New World places.[18] The mechanism of change enacted by the introduction of domesticated animals, Elinor Melville argues, was similar to that enacted by the introduction of diseases. Just as European microbes exploded in "virgin soil epidemics,"[19] so populations of exogenous animals bloomed in "ungulate irruptions"— massive population increases in response to favorable conditions (lack of predation and competition; abundant forage). These dynamics, she argues, were key to understanding postconquest ecological shifts ushered in primarily by *Ovis aries* from Mexico to Australia.[20] Indeed, nowhere were the indigenous plants and animals of the New World more susceptible to recent comers than in Australia and New Zealand, whose histories of geological and evolutionary isolation began approximately 45 million and 80 million years ago, respectively, when each broke off from the great southern supercontinent, Gondwana.[21] Extreme ecological and biological difference—the fact, for instance, that Australia evolved marsupial mammals, and New Zealand evolved none at all, except two species of bat—translated into extreme ecological vulnerability upon the arrival of Europeans and their companion species.

Yet the expansion of Europe since 1500 is not as simple as this genre of neo-Darwinist account would imply. Mere biological advantage did not necessarily translate directly or inevitably into imperial success: making new worlds into neo-Europes required an astounding amount of work, physical *and* intellectual, while ecological transformation was rarely sufficient for lasting conquest without concomitant political and cultural transformations as well. Generations of scholars since Crosby have substantially revised the story of ecological imperialism, adding nuance and layers of complexity to the analysis, not least with respect to the role of domesticated animals in altering the politics of conquest as well as the character of place. In William Cronon's famous account of ecological change in colonial New England, colonists' wandering cattle needed fences, the construction of which ushered in a major shift in land use practices in the area, and ultimately, by provoking conflict between American Indians and early New Englanders, dramatically altered the political as well as the ecological landscape.[22] In Virginia DeJohn Anderson's analysis, similar dynamics are key to understanding early American colonial history more broadly: ideas about, and interactions with, domesticated animals provided a point of distillation in native-newcomer engagement, making the study of colonial domesticates a particularly effective way to gauge more precisely American Indian responses to colonization and conquest.[23] In what is perhaps an even more trenchant revision of standard accounts, recent studies of Britain's pastoral empire in the southern hemisphere have shown convincingly the scale of the effort—literal and metaphorical—required to make Australia and New Zealand into suitable habitat for European domesticates, suggesting that much more went into ecological imperialism than merely differential biological advantages.[24]

And while this work has indeed complicated and enriched our understandings of ecological imperialism and the environmental transformations wrought by the expansion of Europe, almost without exception such studies interrogate the role of animal domesticates in empire at the level of the species. In one sense, this reflects the documentary record left by livestock breeding at the time of Europe's expansion in the early modern period as much as it does the historical conditions governing early importations to the New World. For instance, few details other than numeric ones surrounding the first bovines imported to colonial America exist. William Bradford's *History of Plymouth Plantation* merely lists "3 heifers and a bull, the first begining [sic] of any cattle of that kind in the land."[25] A rare exception to this rule was the importation of Dutch Holsteins to New Amsterdam in 1621, as Deborah Valenze documents in *Milk: A Local and Global History*, although these creatures soon succumbed to the harsh conditions of the mid-Atlantic (and at a faster rate than English varieties).[26] In part, though, it also reflects the fact that a breed as a recognizable distinction within a type is a relatively modern concept, emergent only in the eighteenth century. Fine-grained classifications of domestic animals, as Harriet Ritvo has shown, are both recent and extremely changeable phenomena. Lay and scientific classificatory schema that parsed the diversity of the natural world, including the multitudes of domesticated creatures, were important cognitive components of the imperial world view.[27] The historical record for the nineteenth century better conveys this new (or renewed) significance of distinctions finer than that of the species, a fact increasingly reflected in recent historical scholarship. Specialists in livestock history like Margaret Derry and Robert Peden, for instance, attend to the roles of varieties and types within contained regions of the British Empire (Ontario and the South Island of New Zealand, respectively), and Nathan Sayre's microhistory of ranching and ecological restoration in the Altar Valley of Utah is attuned to distinctions between drought-resistant and other breeds of cattle.[28] Despite these exceptions, scholarly analysis of livestock in empire in the nineteenth century continues to portray the species as the salient analytic.[29] Cattle matter as relatively undifferentiated agents of ecological change, replacing bison in the nineteenth-century American West, for instance;[30] pigs, in William Cronon's definitive account of Chicago's frontier economy, function primarily as a biological storage mechanism for corn on its way to becoming pure capital.[31]

This book, on the contrary, asks what happens if we look beyond species to the level of the *breed*. If we retrain our focus on the breeds of empire, not just in isolated regions of the colonial network, but across Britain's "neo-European" dominions, what new aspects of imperialism and colonialism come into view? To what extent do subspecific distinctions in kind matter to the history of the British Empire? As it turns out, a great deal. Species continued to matter in colonial and imperial experiences in the nineteenth century, in all the ways existing scholarship has shown. But it is at the level of the breed that we see the workings of imperial economy and cultural concerns collide. Focusing on subspecific distinctions is a powerful way to retell a familiar story of changing patterns of production and consumption in the nineteenth century, at both the national and imperial scales. Cattle and sheep were not merely economic animals—their fates were governed as much by cultural imperatives as by those of profit. In the nineteenth century, to colonial producers as much as to British consumers, a sheep was not merely a sheep, nor was a cow or a bull merely that. Different types, crafted over time in response to particular ecological contexts and economic circumstances, evoked different responses on the market, and produced different effects in various colonial places. The view that emerges, therefore, is considerably more complex.

Looking at this level necessitates a change in scale with respect to ecological imperialism. Broad categories—domesticated animals, ungulates, livestock, or even sheep or cattle; politics; land use; and so on—continue to matter, but it becomes apparent that processes of transformation were driven neither by ecological nor by economic concerns alone. Cultural significations, often articulated as matters of taste, were central to the evolution of imperial livestock production in the nineteenth century and its attendant ecological consequences. In aggregate, attention at this level produces a more carefully shaded picture of environment, commerce, and empire than previously understood. The breeds of empire were not blunt tools of domination only, but subtle levers of power, able to take hold of land, to sustain occupation over time and across economic circumstances, and to forge and maintain connections between the imperial center and the colonial peripheries.

Their ability to do so ultimately came down to the perceived relationship between type and place—a relationship that was filtered through the idea of native belonging in the nineteenth century, and that was understood to be highly susceptible to human intervention. What "native" stood for was highly contingent. As a signifier, it was crucial, in political, environmental, and economic terms alike. It was, and remains, a powerful term. Broadly speaking, assigning someone or something to this category could either justify or delegitimize its presence in a given place, but precisely how this played out was far from self-evident, as tracing its usage and meaning across contexts reveals. "Native" was malleable enough to be used for a range of interventions in national and imperial political economy. Its attribution to particular kinds of sheep in the British Isles during the Continental and Napoleonic Wars at the turn of the nineteenth century, for instance, reflected wider concerns about national economic self-reliance, and about Britain's relationship to its European neighbors (the subject of chapter 2). More broadly, in Britain, being "native" intersected with ideas about nationhood and citizenship in ways that influenced what labels like British, Welsh, English, and Scottish meant, and how they were deployed. In the colonies, where Europeans were continually under pressure to legitimize their own presence, the term carried even more weight, especially in the face of indigenous—that is, "native"—peoples whose occupancy evidently predated the arrival of Europeans. With respect to livestock, breeders sought to match particular types of sheep or cattle bred for a specific set of cultural, economic, and environmental circumstances in Great Britain to roughly analogous colonial conditions. In this carefully calibrated process of imperial translocation, the significance of the term shifted—from one designating British origins to one capable of encompassing novel notions of colonial belonging as well.[32]

In the colonies, such ideas about native belonging intersected with understandings of acclimatization—the idea that climate and environment effected lasting change on transposed organisms—in important ways. As the process of biological adaptation to circumstances across (or sometimes even within) generations, acclimatization suggested that people and animals of European descent could, over time, become native to colonial places. While in some ways, the colonial endeavor itself depended on at least the possibility of colonists gaining a native claim to their appropriated lands, in other ways Britons were deeply ambivalent toward such processes of naturalization. Creolization, the perceived cultural and physiological changes that individuals of European descent underwent in the colonies, bore resemblance to ideas about acclimatization and environmental naturalization, but encompassed a broader process of transformation in response to cultural and economic, as well as climatic, conditions. Whether or not the possibility of human acclimatization and adaptation was something to be wished for or feared depended to a great extent on political and cultural circumstances as well as environmental ones. The early phases of colonization, for instance, were often marked by more optimism regarding both Europeans' abilities to adapt and the potentially positive outcomes of creolization than later periods, when the combination of longer experience and complex political imperatives made acclimatization both a riskier and less likely proposition, as Mark Harrison and Joyce Chaplin have argued for India and North America, respectively.[33] Increasingly, over the course of the nineteenth century, the related notions of creolization and acclimatization carried a negative connotation, and the suggestion that any kind of alteration in response to colonial circumstances was a form of deviation, and by implication, degeneration, from a European norm. Significantly, this shift was fed to no small degree by scientific agriculture, where concern over the degeneration of land, and consequently the people who worked and lived on it, particularly in America, was growing.[34]

The stakes of these discourses were highest with respect to people, but how they applied to breeds of livestock in the nineteenth century was by no means inconsequential. Most significantly, new, so-called native colonial breeds supported colonists' claims to foreign lands in the face of existing societies with obvious prior claim: they operated, in effect, as proxies for British people. In situations where human colonists feared the consequences—political, social, moral, and biological—of "going native," British sheep and cattle could do so with relative impunity, thereby bolstering the legitimacy of imperial occupation, the expropriation of indigenous lands, and the disenfranchisement of indigenous peoples. The circulation of British breeds and the concomitant establishment of new "native" colonial breeds are thus crucial to understanding how the question of colonial legitimacy—what is perhaps the central problem in imperialism and its historiography—was worked out on the ground and in the paddock.

Critically, these processes played out according to a logic of consumption: the end game in the dispersal, circulation, formulation, and reformulation of the breeds of empire was meat, meat to feed Britain's industrial population. Thus the material outcome of this discursive and rhetorical conversation about nativeness ultimately hinged on the transformation of a living creature into a commodity. Historians have long understood the significance of commodities—their valuation, circulation, and exchange—to the history of empire.[35] It was mercantilists' appetite for goods and concomitant need for markets, after all, that was a primary cause in the expansion of Europe, according to classic accounts: Columbus and his contemporaries struck out, so the story goes, in search of the valuable commodities of the Far East. When they found the Americas instead, they made do with (and soon profited from) its unanticipated contributions to Europe's economic (and cultural) spheres: land and mineral wealth; timbers and furs; and increasingly the produce of enslaved African labor on agricultural plantations—sugar, tobacco, and eventually cotton.

How such commodities circulated is no less salient—perhaps, even, it is more salient—to understanding empire in the nineteenth century. Scholars of this period have demonstrated that the flows of commodities—from sugar and coffee to soap and cloth—forged and solidified connections between metropole and colony, and laterally between and among colonies themselves. Take sugar and tea, for instance. In Sydney Mintz's classic account, these two imperial products revolutionized English dietary rituals: through their intertwined stimulant and caloric impact, taking tea—that most English of acts—became the metabolic energy of Britain's industrial revolution in the nineteenth century.[36] Moreover, this very act of consumption (and many others like it), as Catherine Hall has argued, brought the empire "home" in ways that made the lived experiences of Britons from London to Aberdeen and nearly everywhere in between inescapably imperial: the goods and materials of empire were themselves constitutive of national identity and of imperial power.[37]

And yet, there is something particular at work in the making of a living creature into a comestible commodity—something, arguably, not at work in the production of sugar or tobacco, cotton or tea. The politics and symbolics that attend this transformation of animal into edible have drawn considerably more ethnographic attention than they have historical scrutiny: Noelie Vialles, for instance, locates in the removal of the act of killing animals for consumption from public view in the nineteenth century a hallmark of the modern era and its particular sensibility.[38] As slaughter receded from sight, the space between animate being and comestible flesh grew, making the ultimately global industrial system of meat production cognitively and emotionally possible, she argues. Subsequent work has refined this central insight. Timothy Pachirat, for example, ably examines how the process of killing in a modern industrial slaughterhouse works to strip individual animals of their distinctive characteristics: size, sex, and degree of curiosity are all erased as body becomes carcass.[39]

For these scholars, the making of meat as a commodity begins at the slaughterhouse gate. Attention to breeds shows, to the contrary, that the process of commodification—that of making meat—begins with reproductive control: the tactics of selection employed by breeders in the nineteenth century were undertaken, always, with the whims of the market in mind.[40] The formulation, replication, and controlled variation of imperial breeds was tuned to the demands of a metropolitan public—one that, moreover, famously invested a great deal of its national identity in the flesh on its tables, as Linda Colley, Steven Shapin, and others have shown.[41] The British since the eighteenth century were nothing if not gustatory chauvinists. That this most British of products—meat—was increasingly bred, raised, and killed abroad has consequences, therefore, for understanding nationalism in an imperial age. If, indeed, we are what we eat, then to be British in the nineteenth century was to be imperial.

Paying attention to the breeds of empire, then, invites us to reconsider our intersecting understandings of empire, environment, and economy. Tracking the production and the shifting signification of "native" breeds in the British Empire in the nineteenth century reveals that the livestock industry of the nineteenth century is not a story about the triumph of economic rationalism, as has often been assumed, nor is it a simple case of commodity circulation or even of ecological substitution: none of these views alone can adequately explain the history of British breeds in the empire. Nor can the story of native British breeds be fully divorced from questions governing human identity, ethnicity, and national or imperial belonging. Debates over the meaning of "native" with respect to sheep and cattle were debates over the degree to which identity was hereditary or territorial, salient questions in the context of the imperial diaspora, human and nonhuman, of the nineteenth century. It *is*, therefore, a story about the generative intersection of culture, commerce, and ecology—about the shifting meanings and uses of native British breeds in the world of imperial livestock breeding.

Topography

As with any history of the British Empire in the nineteenth century, this volume is necessarily selective. Choosing to follow the breeds of empire— and only several breeds at that—means that the geographic coverage of this study is idiosyncratic. Attention to domesticated animals almost necessarily shifts the analytical focus to the "white dominions"—places that were ecologically and climatically enough like conditions in Europe ("neo-Europes," in Crosby's terminology) to be remade in the image of the Old World, and thereby made suitable for European agropastoralism. Still, this work makes no claim to comprehensive study of even this familiar subset of colonial places, and the selective focus adopted here may confound some readers. After all, South Africa is notably absent, while Argentina and the American prairies loom larger than the Canadian dominions. To no small degree this reflects the ways in which these breeds effectively chose their own fates. Hereford cattle went in far greater numbers to the Americas than to South Africa; British breeds of sheep went in far greater numbers to New Zealand and Australia than to South Africa or North America.[42] They did so initially at the hands of the human beings who transposed them, but whether or not they became established in new homes depended on whether and how they acclimatized to colonial conditions—a process determined as much by the breed as by the breeder. Thus the contours of this study reflect the contours of the world of these breeds as it was in the nineteenth century. These match, in some places, the boundaries of the British Empire. In others, they push beyond it, reflecting the more amorphous sphere of Anglo influence sometimes called "informal" empire, and sometimes called the "Anglo World." As such, this study contributes to a growing willingness in environmental and global history to look past national boundaries, and in particular, to reintegrate the history of the United States into imperial and environmental history more broadly.[43] Here, the American West is treated as one among several "neo-Europes," as much a part of Britain's sphere of influence as Argentina or Uruguay.

The layout of the chapters reflects this orientation, as it reflects the origins, the circulation, and ultimately the return of native British breeds in the empire, in one form or another. Part 1 is concerned with the origin and formation of "native" British types. Chapter 1 explores the growing recognition of "native" British breeds of sheep and cattle at the turn of the nineteenth century, arguing that it was a somewhat unexpected outcome of the mania for breed improvement that marked the eighteenth century's agricultural revolution. In the late eighteenth century, breeders enthusiastically relocated various types of sheep and cattle from one part of the United Kingdom to another, prompting them to reflect on the degree to which a seemingly fixed type was susceptible to the effects of human intervention as well as environmental considerations. This activity gathered pace in the third quarter of the eighteenth century as a spirit of improvement motivated inquiry into all matters agricultural, including the herds and flocks of the British Isles. Agricultural reportage stimulated interest in, and recognition of, the great variety of breeds in different localities throughout England, Scotland, and Wales at just the time that enterprising breeders were "improving" their stock, selecting for early maturity and high yields. In the enthusiasm that ensued for these novel animals, the influence of these types spread far and wide as breeders crossed their existing stock with improved varieties like Robert Bakewell's New Leicester Longwool. Combined with a form of patriotism that drew explicit links between agricultural production and the glory of Great Britain, these efforts, though controversial and to some extent limited by the constraints of climate and ecology, helped to lay the foundations for a national taste for British meat.

Chapter 2 picks up the meaning of "native" breed in the context of the failed attempts to naturalize the Spanish merino in Great Britain at the turn of nineteenth century. As breeds were standardized in the early nineteenth century, sheep breeders increasingly responded to demand for more and more varied types of wool, and especially to growing demand for very fine fibers that could be woven into soft, luxurious cloth. Without a doubt, the best breed of sheep for producing such fine wool was the merino, whose origins lay with the ancient pastoral economy of Spain. Consequently, agricultural improvers, under the aegis of King George III himself (known as Farmer George for his enthusiasm for agricultural pursuits) avidly promoted the Spanish breed's adoption in Britain.[44] But producing merino wool in England was not as straightforward as simply raising merino sheep on English soil and waiting for the fleece to grow.

Merino sheep are extremophiles, lovers of scorching heat or freezing cold, but of little in between. While they thrived as introductions to Great Britain's recently acquired Australian colonies during this same period, they were almost spectacularly ill-suited to the relative mildness and perpetual damp of the British Isles. Efforts to acclimatize them there ran up against the limits of physiology and the effects of climate, but they also offended the sensibilities of dyed-in-the-wool champions of English "native" breeds. Detractors railed against the merino's unsightly form and unappealing foreign flesh, apparently anathema to the British taste in meat—while its proponents defended merino mutton, declaring their preference to be more refined than that of the uncouth supporters of dedicated mutton breeds. Ultimately, champions of the Spanish breed lost on both fronts, culturally *and* economically, as merino wool quite literally deteriorated in the British Isles, becoming heavier and coarser in the hands of British breeders, perhaps partly in defense against the damp and rain, while the breed's stubborn foreign carcass withstood the best efforts to transform its alien aspects. Efforts to produce an English sheep in Spanish clothing failed. The tastes of Britons and the climate of their island home prevailed. The fate of the merino in Great Britain became an object-lesson on the power of climate and the limits of human virtuosity.

Chapter 3 pursues the idea of a native breed in the context of pedigree cattle-breeding in the first half of the nineteenth century, by examining its changing significance for the Hereford breed of cattle. Named for their native home of Herefordshire, situated on the Welsh border, the Hereford proved remarkably able to thrive across a range of environmental conditions, systems of production, and market imperatives. Initially it was known as a hardy, multipurpose breed accustomed to spending years in front of the plow before making its way to the butcher's, and until the middle of the nineteenth century, a rather uncouth variety, especially compared with such improved breeds as the Shorthorn.[45] But with the ascent of techniques of improvement—which included pedigree record-keeping, published herd books, and official breed societies, as well as careful selection and, in this case, intensive inbreeding—"purity" became the watchword of cattle-breeding.

Against the "improved" Shorthorn, the benchmark for nearly all things bovine in the nineteenth century, the Hereford now seemed wanting. Its native connection to the county of Herefordshire became a way for proponents of the breed to claim purity of descent in absence of an official herd book, through geographic localization and the presumed connection to antiquity that it conferred. When this metric proved insufficient, homogenizing the breed's phenotype, which historically had varied from speckled to dove-gray to all red, into a white-faced, red-bodied type, became a visual signifier of consanguinity. These metrics were necessarily artificial and demonstrably fabricated, but this did not detract from their utility. It merely reflected the artificial and fabricated nature of purity itself as a desired end, and not even the pedigree—the published genealogy of any "thoroughbred" animal—could ensure the desideratum against its own illusory character.

As Herefords became more refined, they increasingly shed their identity as a "native" county breed. As they, and other types of cattle and sheep, began to be exported from Great Britain in greater and greater volume, these erstwhile local varieties also took on the label of "British." With the expanded scope for breeding and production that the imperial domain presented, "British" became an important qualifier for breeds of livestock, not only denoting geographical origins but also, as stock breeders would have it, serving as a guarantee of excellence. Within the British Isles, too, this kind of signification was important. With growing prosperity in the nineteenth century came an increase in the British national appetite for meat, and Herefords—along with several other breeds—became synonymous with the "roast beef of Old England," a supposed culinary tradition whose significance to the social, political, and environmental histories of Great Britain (and, indeed, the British Empire) cannot be underestimated. At the same time English, Scottish, and Welsh remained important signifiers, although, like British, ones that were increasingly employed in reference to type rather than to location. They came to indicate particular attributes (like hardiness, in the case of Welsh, or meatiness, as was often the case for Scottish), and they remained especially relevant on the market and at the butchers where they delineated important distinctions in value. As the middle class expanded, it became increasingly important to educate people's tastes in all areas of cultural and commercial consumption. The British nation's taste for beef was produced at the same time that breeds like the Hereford were "improved," the two phenomena (taste and breed) developing in tandem, as commercial and social factors, particularly increasingly sophisticated merchandising, and cheapening costs of production contributed to the rise of the Sunday roast as a venerable institution.

Part 2 moves beyond the shores of the British Isles as native British breeds began to circulate widely throughout the empire, and in some cases, beyond. By the third quarter of the nineteenth century, questions about the degree to which environment, climate, and now national origins could be embodied in a breed were now playing out in the colonies. Chapter 4 explores these dynamics in New Zealand, where *O. aries* was the dominant species in the colonial pastoral economy. With neighboring Australia, it was an important producer of wool for British manufacturing, and until the 1890s, merino sheep predominated throughout Australasia. As global wool prices fluctuated with dramatic intensity from the 1860s on, though, producers soon sought to diversify the products they could extract from their flocks. The ovine population of New Zealand far outnumbered potential local human consumers, and this problem of surplus meat only intensified as profits from wool plummeted. Colonial breeders became desperate to find an outlet for their mutton. Refrigeration technology, coupled with steam-powered shipping, offered a way out of this economic impasse. It meant that Australasia's ovine surplus could be sent, in frozen form, to satisfy the demand for meat in Great Britain, whose appetite, with its population, had been growing throughout the third quarter of the nineteenth century.[46] Together these factors spurred the development of new breeds in New Zealand, colonial hybrids that bridged the realities of environmental and economic conditions in the colonies, and the demands of British consumers, whose preference for British meat had changed little since the rejection of merino mutton in the 1810s.

In this, superficial resonance with the British Isles in climate and topography served New Zealand well. These temperate islands of the South Pacific resembled the British Isles to a certain extent, making the transposition of British meat breeds like Lincoln Longwools, Southdowns, and Romney Marsh sheep a success, but native flora and topography, as well as rudimentary transportation networks and the entrenched wool economy, complicated growing British mutton on New Zealand's soils.[47] A large proportion of the colony's flocks were to some degree merino. And New Zealand's climate was far more extreme than the temperate British Isles. Strong winds buffeted its steep hills, the heat and sun of summer as well as the chill of winter often far more intense than England's. Breeders had to battle against their flocks' tendencies to adapt to these conditions if they wanted their meat to remain British. In this fight, steam technology was their ally, the constant contact permitted by the networks of oceanic steam transport enabling them to resist the uncontrolled creolization of their sheep with regular infusions of breeding stock from Great Britain.[48]

The access to British metropolitan markets that this technology granted also shaped the development of colonial sheep breeds, which needed to satisfy the twofold (and sometimes contradictory) requirements of the climates and topographies of Australasia, on the one hand, and the tastes of British consumers, on the other. The solution was to create colonial breeds out of existing stock of British and merino extraction by crossing the rams of heavy mutton breeds like Lincoln Longwools or Romney Marsh sheep with merino ewes, and then rigorously inbreeding successive generations for desired characteristics. The desideratum in this endeavor was the combination of fine wool and meaty carcass—the very blend of traits that had so frustratingly eluded naturalizers of the merino sixty years earlier in Great Britain. Unlike in the British Isles, the climate of New Zealand aided colonial breeders, and the outcome was the production of both "new" breeds and colonial versions of existing ones. For instance, the Corriedale, born primarily of a Lincoln-merino cross and especially suited to the expansive, dry grasslands of New Zealand's South Island, was promoted as "New Zealand's own,"[49] while "New Zealand Romneys" retained discursive and physiological resemblance to their parent breed, producing a fleece with much of the fineness of merino wool, but maintaining the characteristic resistance to the ill effects of wet land developed in Lincolnshire and the waterlogged fens in the east of England—a trait that proved useful in the more humid, coastal plains of the South Island and the subtropical reaches of the North Island. Such breeds embodied the tension between colonial environments and imperial demands: their hybridity guaranteed suitability for colonial topography and terrain, while their genetic roots ensured that they remained British enough for metropolitan consumers.

Chapter 5 addresses the question of how to manage nativeness overseas from a different angle, through the case of Hereford cattle, which were widely exported in the 1870s. A corollary of concerted efforts to feed Britons at home with quality (i.e., British) meat raised abroad, Hereford cattle, especially bulls, were exported in great numbers to North and South America, as well as Australia and New Zealand. In their enthusiasm to profit from the demand for high-quality animals, which overseas ranchers used to improve the quality of their existing herds, breeders in Britain enabled their (quasi-)colonial competitors to establish their own high-quality reservoirs of the breed's genetic potential. Conflict ensued in the 1880s over the relative stringency, or lack thereof, in English pedigrees, which effectively cut the ties between the North American and British breeders and their bloodlines. An insult to the pride of English Hereford breeders, it meant that henceforth, the two branches of the family tree were bred in near isolation. Enthusiasm for native breeds as a way to produce British meat absent British pasture ultimately jeopardized the embodied hereditary connection between American and British Herefords. Much like the Corriedale in New Zealand, the former became a new, quasi-colonial "native" breed.

This set the stage for the invention of "Traditional" Hereford cattle. With the widespread enthusiasm for novel imported Continental and foreign breeds in postwar Britain came a growing recognition of the merits of these erstwhile colonial cousins of the English Hereford. Beginning in the 1970s, foreign-bred Herefords began to be imported to Britain with great enthusiasm.[50] Reared primarily in Canada but also in the United States and Australia, to thrive under the highly routinized conditions of modern industrial meat production and its notorious system of feedlots, these formerly colonial cattle were taller, meatier, and faster to reach maturity than the short-legged and hardy but slow-maturing English Hereford. As these cattle were crossbred with these exogenous types, some breeders began to notice a change in their preferred variety, and worried that it was in danger of complete obliteration under the tidal wave of American and Canadian Herefords. Proponents of the English type therefore took steps to defend their preferred, "pure English" type—the Traditional Hereford: those cattle whose entire lineage could be traced to individuals bred only in the British Isles. In so doing, they implicitly privileged environmental factors over shared genetic roots: time spent outside Britain and in the hands of unfamiliar breeders, not common origins in nineteenth-century British stock, came to define the reimported former colonial varieties.

These anxieties coincided not only with the disintegration of the former empire but also with the successive waves of immigration of former colonial subjects to Great Britain.[51] Concern over the impact of postcolonial creole breeds upon native British breeds of cattle mirrored disquietude over the effects of widespread human immigration on British society and culture.[52] As the more racially and culturally diverse population of Britain evolved in the late twentieth century, the realm of breed conservation remained an unusual discursive space in which conversations about English purity and nativeness continued to take place unapologetically. Defending native breeds like the Traditional Hereford against colonial upstarts returned to their erstwhile native land privileged generations spent on foreign lands in unfamiliar climates as defining factors, rather than shared genetic and historical roots. In its reluctance to acknowledge the legitimacy of these animals in Great Britain, such a position also denied imperialism as a reciprocal process in which the creole formations of the colonies came "home" to roost.

CLAIMS ON BEHALF of a breed to nativeness are thus never simple, nor have they ever been innocent. Such rhetoric always transcended the merely geographical, and spoke to political imperatives, to a concern for status, to ecological anxiety. The pliability of the concept ensured its long shelf-life: for each case, at each time, in each setting, what "native" signified was redirected; what it included or excluded was reinscribed. Charting its changing meaning reveals the inner workings of colonialism—the material and figurative production and sustenance of ties between colony and metropole which, in the cases explored here, took "the shape of meat."[53] Doing so also reveals the cost of the agropastoral imperial endeavor. Following the invention, reinvention, and transformation of British breeds from cradle to grave—from home county, to distant land, to icy tomb, and eventually to British table—exposes the foundations of a globalizing industrial apparatus for meat-making, one that profoundly influenced the shape of distant societies and ecologies, and whose legacy supplies the tables of the developed world to this day. Enabled by the modernized, standardized British breeds that emerged from a national mania for agricultural "improvement" at the end of the eighteenth century, the cost of this system of production has been the concurrent standardization of place—the erosion of local distinction and biotic variability the world over, the consequences of which we continue to wrestle with today.

PART I | Great Britain

CHAPTER ONE

A Breed in Any Other Place

When William Brown, the author of a "handy-book to the science and practice of British sheep farming," described the geography of the principal breeds of sheep in the British Isles in 1870, he estimated that "no less than 63 per cent of the British Isles [was] still claimed by the native breeds."[1] The remainder was the territory, in his calculus, of the other dominant types of British sheep: Cheviots, the improved hillside breed that dominated Scotland and the north of England; Down sheep, originally adapted for the chalky hills and vales of southern England; and Leicesters, that most highly cultivated kind of British ovine, the creation of the esteemed Robert Bakewell (1725–1795), and the darling of late-eighteenth-century improvers. Unlike these well-defined and carefully honed breeds, the "natives" to which Brown referred were a motley crew, including the Blackfaced variety—a common designation that was itself a catchall category for various kinds of mountain sheep in England, Scotland, and Wales—and Irish sheep, a class about which "little," Brown noted ruefully, "can be said in [their] favour."[2]

Of course, there were many more breeds of sheep in the British Isles in the 1870s than these four, a fact of which Brown was well aware. Indeed, diversity of type was the hallmark of British animal husbandry. Nearly each region, each locality, boasted its own distinctive type of cattle or sheep, and it was these local kinds—many of which remained largely unaltered by the techniques of "improvement" that dominated British and Scottish agriculture from the late eighteenth century until well into the nineteenth—that Brown meant to encompass within the category of native. For Brown, this designation affirmed the effects of nature over those of human ingenuity or artifice when it came to shaping and determining type. Unlike Bakewell's New Leicester breed (one "purely of man's modelling,"[3] as the author put it), Blackfaced mountain breeds and other native kinds had "had no handling except in careful selection among themselves."[4] Their form, according to Brown, had been determined entirely in response to their surrounding conditions, to the cumulative effects of competition, predation, scarcity, and sexual selection—what Charles Darwin, with the publication of *The Origin of Species* in 1859, had recently given wide public and scientific currency to as the processes of natural selection.

Map of the British Isles, "Exhibiting [the] Distribution of Prevailing Kinds of Sheep." From William Brown, *British Sheep Farming*.

What it meant to call a type of livestock "native" to a particular place or environment varied considerably according to temporal, geographic, and political context. Indeed, it is the task of subsequent chapters to trace some of the situational variations in the idea of native breeds, as well as their political, cultural, and economic consequences, both within Great Britain and in parts of its former empire. But despite the wide array of meanings that could be—and were—attached to the term over the course of the nineteenth century (and beyond it), using it as Brown did to differentiate between highly cultivated improved types on the one hand and unimproved or local varieties on the other remained a consistent usage over time and across context. The predominance of this denotation, however, did not prevent other meanings from adhering to the word "native," which was always multivalent when applied to livestock. Nor were the political implications and cultural, as well as economic, valuations that spun out from the idea of nativeness necessarily stable or narrowly construed: they could (and did) proliferate, varying according to time, place, and context. And yet, at a foundational level, calling a breed "native" worked to dissociate it, consistently and persistently, from notions of refinement, from cultivation, from purposive human design. This chapter explores the origins of the term as applied to breeds of sheep and cattle, arguing that this basic lexical stability can be traced to the "discovery" of native breeds during Great Britain's fit of agricultural improvement in the late eighteenth century, and to the constantly shifting assumed roles for heredity, environment, and human selective pressure in understandings of a breed itself. Which of the latter of these—environment and climate, or anthropogenic selection—had the most significant impact over the former—the transmission of traits from parent to offspring—effectively shaped how agriculturalists at the turn of the century understood the notion of a breed. And while eighteenth- and nineteenth-century stockbreeders operated on the basis of a largely tacit body of knowledge, and in the absence of a convincing understanding of the principles or mechanisms of heredity (these emerged slowly, in piecemeal fashion, and amid much contention between the 1890s and 1920s) they were adept at recognizing and managing the effects of both inheritance and environment.

Every Soil Has Its Stock

Today it is understood that as the sum of flocks or herds, themselves aggregates of individuals, a breed is by definition subject to constant change.[5] Animals are born, reproduce, and die. In the process, some—but not all—traits or characteristics are transmitted across generations, and therefore the appearance, instincts, and even the behavior of particular individuals are liable to diverge from that of their fellows or forebears. How a given population looks or behaves—what it eats or how it reproduces, where it migrates or the range it is able to inhabit—are thus all subject to change over the course of generations. This means that a breed is an inherently unstable thing, the word itself denoting a sense of transmission absent in synonymous terms such as kind, type, variety, strain, or race. It came into wide currency "among husbandmen" only in the early nineteenth century as a way to distinguish between "varieties [of stock], possessed of peculiar characters" precisely because its relation to the verb *to breed* suggested the means by which "it is supposed their respective properties are in great measure communicable to their descendants."[6]

Never entirely containable by the methods of selective breeders, any breed's genome contains enough variation that individual phenotypes will differ considerably. Even in a "closed" breed—inbred for many generations to produce homogeneous genotypes in its composite individuals, and therefore a relatively stable phenotype—genetic drift sufficient to modify the overall characteristics and appearance of the breed is the norm. Indeed, the purebreeding techniques of the eighteenth and nineteenth centuries, in which consanguineous individuals (including, often, parents and their offspring) were bred to each other, worked to forestall as much as possible this inevitability by concentrating certain qualities, in contemporary parlance, or honing the genotype, in modern terms. The peril of purebreeding was that such pairings could concentrate undesirable traits as well as desirable ones, but it was nevertheless the preferred approach to stockbreeding, and a necessary one. As John Sebright explained in 1809 in an influential pamphlet on breeding, "Breeding *in-and-in*, will, of course, have the same effect in strengthening the good, as the bad properties, and may be beneficial, if not carried too far, particularly in fixing any variety which may be thought valuable."[7]

This tendency to variability—one common to all forms of life—though, could also be an advantage to livestock breeders. It was, in fact, the material starting point for selective breeding. "Individual variety of size and shape prevails in all breeds," John Lawrence—an expert on the "breeding, management, [and] improvement of domestic livestock"—wrote in 1805, "to the infinite use and convenience of man."[8] Or, as Charles Darwin put it in the introduction to the first volume of *Variation of Animals and Plants under Domestication* (1868), "If organic beings had not possessed an inherent tendency to vary, man could have done nothing."[9] This innate and inevitable tendency was crucial. It produced a gap between heritability and physical appearance—between genotype and phenotype—that offered breeders, as it were, the space to experiment with selection. "Although man does not cause variability, and cannot even prevent it," Darwin wrote, "he can select, preserve, and accumulate the variations given to him by the hand of nature almost in any way which he chooses," and in this way, the careful breeder of livestock can thereby "certainly produce a great result," whether this was the production of long or fine wool, rich or copious milk, fine hides or fatty meat, or some combination thereof.[10]

Beginning in the late eighteenth century, this was increasingly the territory of enthusiastic agriculturalists working under the sway of an ethos of "improvement." As a guiding philosophy, improvement was applied to many branches of rural economy—to agricultural technology and farming implements, crop yields, the study of soil content and quality, to large-scale environmental engineering projects of draining and irrigation, and even to the climates of northern locales in an effort to ameliorate them[11]—with the overall goal of rationalizing, and thereby increasing the productivity of, the agricultural sector of Great Britain. It resulted in several major developments, not least among them the transition from common grazing to enclosed pasture and cropland, a sweeping change to the landscape of the British Isles, with devastating social consequences, especially in Scotland.[12] More prosaically, but perhaps no less revolutionary in the long run, agricultural improvement also gave Britain the rutabaga, or Swede: originally a humble Swedish vegetable, it dramatically transformed livestock feeding and fattening.[13] And it led to the widespread adoption of the four-field system of crop rotation, in which "the cultivation of clover and rye-grass, joined to . . . turnip husbandry" and to the pasturing of livestock, resulted in "such luxuriant crops of grain . . . as could not be produced, by any other means."[14] Improvers, many of whom hailed from the landed aristocracy, promoted such efficiency at all scales, from the smallest holding to the greatest estate (although they continually bemoaned the common farmer's resistance to such activity), efforts they undertook in the name of enhancing the prestige along with the material well-being of Great Britain.

Agricultural improvers turned their eye toward the nation's hordes of sheep and cattle, too. Under the gaze of improvers like John Sinclair, John Southey Somerville, Arthur Young, and others, the regionally, even locally, distinguishable breeds that predominated in England, Scotland, and Wales in the late eighteenth century increasingly came into focus. In fact, the modern notion of a "breed" as a replicable type itself gained currency around the same time in recognition of two concurrent foci in the shaping of livestock—the ability of a skillful breeder to impress human desiderata (size, color, form) on a group of animals, and increasing awareness of the variety and distinctiveness of type throughout the British Isles.[15] Acknowledgment of both arose in consequence of the spirit of improvement that marked agricultural pursuits at this time, the former from the practical application of this philosophy to livestock production, and the latter from the movement's attendant desire to promote useful agricultural knowledge. Enthusiasm for useful knowledge, which of course extended well beyond the agricultural sector, and only increased over the course of the nineteenth century, spurred the growth of a genre of agricultural reportage describing rural affairs throughout England, Wales, Scotland, and eventually Ireland, designed to encourage enlightened practice among the nation's landowning class.[16]

In addition to accounts of geological and climatological conditions, farming economy, statistical values relating to arable and pastoral production, and detailed descriptions of the gardens, houses, farm buildings, and cottages of great estates, the "general views," commissioned by Great Britain's Board of Agriculture, contained details of the particulars of a region's livestock. In aggregate, these accounts of rural economy and animal husbandry offer a picture of a system of livestock breeding at the commencement of the nineteenth century remarkable for its diversity. Each region contained its own, usually eponymous, type of cattle and/or sheep. Breeds were called after their particular localities not merely from convenience or because that was where they could be found, but because it was assumed that the character of a place infused the character of a breed. The cattle of Cambridgeshire, for example, were "mostly the horned breed of the county, and are called by its name," according to William Gooch,[17] while in the southwest of England, Devonshire cattle predominated, "and this breed, more or less pure, prevails throughout Cornwall," according to George B. Worgan, author of the *General View of the Agriculture of the County of Cornwall*.[18] Gradations in type were common, particularly so if the area in question was large, like the Southwest of England. Writing in 1807, Worgan carefully distinguished the fine differentiations within the Devonshire type, North Devons (the preferred breed of "the more enlightened and spirited breeders") being fine and somewhat delicate, while South Devons, "more of a brown, than of a blood-red colour" were "considered stronger and more hardy."[19] Breed, or type, was understood to be intimately connected to the nature of a place. In a region of variable geology and climate like Cornwall, where a "great diversity of soil prevails . . . as well as difference of situation in regard to shelter and exposure, it is not to be wondered at," wrote Worgan, "that the cattle which are bred and fed thereon, should also vary much in size and other properties, occasioned by local circumstances."[20] Indeed, that "every soil has its own stock" was the presiding understanding of the differentiation of breeds within Britain at the turn of the nineteenth century.[21]

The emergence of modern genetics and developments in the field of animal science in the twentieth century have given subsequent generations of theorists and practitioners the conceptual tools with which to develop precise understandings of inheritance, and the role of environment in selection, both artificial and natural. It is now generally conceded that the influences of heredity predominate over the influences of a given individual's environment, and by extension, over those of a population.[22] But because the operation of heritability remained a puzzle throughout most of the nineteenth century, no such consensus then reigned. Consequently, a breed's relation to the influence of environmental surroundings—climate, herbage, soil type, seasonal variation, and so forth—remained open to debate, and the relative contributions of climate and environment, on the one hand, and heredity and selection on the other, were difficult to tease apart.

It was clear to all with knowledge of the subject that inheritance was central to the breeding of superior animals. But it was clear, too, that external conditions played their own role in shaping type. The sheer variety of livestock populating the British Isles lent weight to the supposition that the character of a place infused the character of a breed. Great diversity reigned across British pastures, from the shaggy Kyloe beasts that grazed the inhospitable moors of the Scottish Highlands or the old Longhorn type of Lincolnshire in the northeast of England, to the robust and beefy Hereford, or the highly refined Shorthorn type. Nor was such variety limited to the bovine species. The simplified four-part typology of sheep that William Brown mapped onto the British Isles in 1870 masked a wider array of ovine types including the rare and peculiar (the Herdwick sheep of the Lake District, for instance, which were celebrated for their unusual homing instincts; or the singular breed of North Ronaldsay in Scotland, which evolved to forage for seaweed at low tide), as well as the relatively widespread and more familiar. The fine-wooled Ryeland sheep of Herefordshire; the large and rawboned breed of Lincolnshire; the multihorned and multicolored Shetland breed; the fat Wiltshires, Shropshires, Dorsetshires, and Southdowns; the long-legged and long-wooled Romney Marsh sheep—each distinctive place could claim its own distinctive breeds.

Sheep breeds characteristic of the "Higher Welsh Mountains," showing various degrees of "improvement." From David Low, *Breeds of the Domestic Animals of the British Islands*, Volume 2, Plate 2. Rare Book & Manuscript Library, Columbia University in the City of New York.

Indeed, so evident did this connection between environment and type seem that no one "who has ever had an opportunity of considering the subject," opined John Sinclair, a leading agriculturalist of the late eighteenth century, would "ever entertain the idea, that only one breed of sheep, ought to be propagated in these kingdoms."[23] This sentiment held particularly true for sheep, a species that, at least in the British Isles, seemed subject to an especially high degree of variability. Though "all animals are subject to variety as determined mainly by breed and climate," Brown claimed, "no kind of animal varies so much as the *Ovis* in adapting itself to circumstances, or becoming acclimatised."[24] As Sinclair put it, the "hardy and plastic nature of the animal itself" was a match for the "variety of ground on which it may be safely pastured," lending evidence, in his view, "that nature intended, that there should be a considerable diversity of breeds, even in the same individual country."[25] According to Brown, recognition of this provided "the great starting-point in sheep-culture."[26] And by the time at which he was active, thanks in no small part to the explorative efforts of earlier generations of improvers, the idea that the different "habitats" boasted "prevalent breeds of sheep adapted to them" was common knowledge.[27] "Most people," he wrote, "have an indefinite general knowledge on this question; they have often heard it spoken of in an incidental way, and they know ... that a Down [sheep] will not thrive on the Grampians," even if the "particular reason" for this eluded comprehension.[28] Although the precise "influence of climature, on the constitution, or changeable part of the nature of animals [was] a matter of difficulty to be demonstrated," as an authority on rural economy writing in the late eighteenth century realized, its effects were manifest. "No man has yet been able to breed Arabian horses, in England," this author continued, nor "English horses, in France or Germany; nor Yorkshire horses in any other District of England."[29]

But just as the variety of native types of sheep and cattle could be marshaled as evidence for the effect of climate and environment upon breeds of animals, so too—paradoxically—could it be claimed by those who argued in support of the primacy of inheritance over external conditions. This was particularly apparent when it came to the various qualities of wool grown in the British Isles. "Although we should be disposed to attribute ever so much to the influence of the climate of G. Britain," the author of a 1774 essay on "the improvement of the Highlands" remarked, "yet the difference between the heat of the climate of different places in this island that afford wool of very different qualities, is but very inconsiderable."[30] The diversity of breeds in Great Britain was impressive; its climatic range less so. Not only do conditions and temperatures vary relatively little from place to place, seasonal fluctuations are also limited. With the exception of pockets of extremity—the west coast of Scotland, for instance, or the Peak District in Yorkshire—the British Isles are consistently humid and temperate nearly everywhere. Given this stability, climatic variance alone could not account for the wide variety found among British sheep (and cattle). "Many parts of England enjoy a climate nearly similar to that of Herefordshire and Gloucester," this author noted, "yet wool of an equal quality is not to be met with there."[31] It was not climatic differences or the nature of the pasturage but the "very great diversity that we find in the nature of sheep," he claimed, that accounted for the varieties of wool grown in the British Isles.[32] When it came to understanding the relationship between types of livestock and the places in which they could be found, the evidence, it is clear, remained open to interpretation.

The Eye of the Breeder

If the evidence itself was open to debate, the grounds upon which it was produced were less so. Overwhelmingly throughout the nineteenth century, claims to authority and expertise on what made a breed a breed were made upon the basis of experience combined with careful observation. Before it was codified as genetics or sanctified as a scientific field, knowledge of how characteristics were transmitted from generation to generation operated tacitly.[33] The stuff of livestock breeding was an applied knowledge, a kind of barnyard science, and therefore difficult to transmit in the absence of practical experience. There were limits to what could be conveyed by pen and ink: knowledge gained "by an accumulation of circumstances—ordinarily called experience"—was paramount.[34] William Brown believed that "any amount of reading without the long daily experience of the grazier [was] of little service to the young husbandman."[35] Others concurred. In 1875, a columnist for the *Livestock Journal and Fancier's Gazette*, a weekly publication devoted to the breeding and rearing of pedigreed livestock, poultry, and pets, expressed frustration over the incapacity of "fanciers and breeders... for *appropriating knowledge*" of practical breeding from the pages of the journal, given that such knowledge "cannot be taught in words."[36] Rather than prompting a crisis of professional identity (for what good was a journal devoted to livestock breeding if such knowledge was intransmissible by ink and paper?), this lamentable observation spurred the writer to ruminate on the nature of the "art" of breeding: not an "instinctive" one, it was "simply the result of experience ... constantly accumulated and gathered up to be applied throughout succeeding seasons," and the "thinking breeder" was therefore "ever keeping his own eyes open to apply what he has learnt from others to his own experience."[37] By these means—patience, observation, experience—was "*mastery* of [this] ... particular branch" of knowledge assumed in the absence of what one historian recently called "a functional explanation of biological inheritance."[38]

Experience on the ground, so to speak, and keen observation, then, were deemed critical to breeding of any kind, improved or otherwise. In the debates that marked livestock breeding throughout the eighteenth and nineteenth centuries, these two components remained the benchmark for knowledge and the basis of authority. John Little, author of *Practical Observations on the Improvement and Management of Mountain Sheep, and Sheep Farms* (1815), found the courage to publish his views on "the intricacy of this very interesting subject" based on his thirty years' experience as a shepherd and sheep farmer.[39] Most other authors on the subject of "mountain sheep-farming" "were not practical men," he noted in his preface, alluding in all probability to the fact that the backers of agricultural "improvement" largely hailed from among the wealthy, landed, elite. They rarely got their boots, much less their hands, dirty, preferring instead to delegate the practical work of selective breeding, rearing, and general farm management to a team of practical experts (like Little), overseen by a paid steward.

Consequently, Little claimed, the information their works contained, and the rules they had "laid down ... for the management of every kind of mountain stock, and of mountain farms however situated with respect to soil and climate," were not based on firsthand knowledge.[40] By contrast, Little's own volume was founded, he claimed, entirely upon "the experience I have had in different parts of Scotland and in Wales."[41] This gave him the authority to publish eight chapters on topics ranging from the relative merits and description of the Cheviot breed, to the life of a shepherd, to "observations on the improvement of the country by means of railways."[42] And if, he eventually concluded, he had "fallen into any [errors]" along the way, he was careful to note that it was not because he had "indulg[ed] any favourite theory, but entirely from the inaccuracy of the observations I have made in the course of my practice as a shepherd."[43]

The emphasis that experts placed on the weight of experience helped ensure that the relationship between the formal sciences and the practice of breeding in the nineteenth century was a distant one.[44] Those familiar with both the workings of scientific inquiry and the breeding of livestock agreed that even proximal sciences like physiology and anatomy offered little recourse to practitioners. For instance, John Hunt, a physician and a great livestock enthusiast (though not a practical breeder) who claimed Erasmus Darwin (1731–1802)—the well-known botanist and eventual, albeit posthumous, grandfather to Charles—as his "learned friend," wrote in 1812 that "the breeding and feeding of domestic animals [was] not to be explained" by the "parade of philosophy," but rather by "a knowledge of nature."[45]

But it was not merely that formal scientific theory appeared to have little bearing on the practice of breeding. Breeders, too, tended to overlook developments in fields like natural history that might have furthered their own aims. This remained the case for the most part even after what we now consider to be some of the most significant developments in what became the disciplines of biology and genetics. The wide acceptance of Charles Darwin's theory of natural selection, for example, and his attendant theories of artificial selection, had surprisingly little impact (at least from the modern perspective) on the work of stockbreeders in the latter decades of the nineteenth century. In 1902, for instance, the *New Zealand Farmer*, the dominion's premier agricultural publication, quoted an unnamed writer for an unspecified "English agricultural journal" who lamented the fact that most sheep breeders remained ignorant of the principles of artificial selection that Darwin had laid out in *Variation* some thirty-four years earlier. "So many years have passed since Darwin has published his two-volumed 'Animals and Plants under Domestication,'" he wrote, "that it should have been impossible for any ewe in any well-to-do farmer's flock to produce less than twins except as a rare and unwelcome throw-back to earlier habit." That evolution tended toward multiple births he took for granted, and in failing "to follow Nature's plan and eliminate the less fertile generation after generation," breeders had hampered the progress of ovine evolution and stunted the wealth and prosperity of British agriculture. "If the matter had been scientifically attended to," he claimed, "our sheep should now be passing into the triplet-bearing stage of evolution, fully provided with stamina and milk-producing powers to keep pace with such increased fertility."[46] While ewes bearing litters of lambs may seem a far-fetched evolutionary eventuality, even with the optimism and unquestioning faith in the human capability to mold nature that this writer displayed, the gulf between the science of inheritance and the practice of livestock breeding was a real one. Already apparent in the early nineteenth century, the distance between the two widened and persisted until well into the twentieth century, as Margaret Derry has shown in her recent survey of the science and practice of breeding since 1700.[47]

Even so, this robust body of knowledge founded on practical experience meant that collectively breeders understood the "communication of qualities" across generations very well.[48] If the innate variability of living creatures was the material starting point for selective breeding, the "general rule [that] the young resemble their parents in form and properties" was the conceptual one.[49] Simply put, "in the common phrase," that "like produces like" was the well-known maxim of livestock breeding.[50] "Blood" served as a metaphor for the means and substance of heritability, though contemporaries were aware that it was "nothing more than an abstract term," as one theorist put it, "expressive of certain external visible forms which, from experience, we infer to be inseparably connected with those excellencies which we most covet."[51] That this link between external excellencies and heritability remained largely unsubstantiated was less troubling to breeders and practitioners than might be assumed. For most, it was enough that their tactics worked: the proof, so to speak, was in the progeny, which were demonstrably fatter or woolier, or better milkers, after several generations of selective breeding.[52]

Because heritability was necessarily an embodied phenomenon, the ability to discern particular traits, qualities, or characteristics of certain animals—such as early maturity, fattening ability, a certain bone structure, a generous fleece—was paramount. "Intelligent breeders are now aware," wrote Andrew Coventry, an early authority on breeds and breeding, "that the different kinds of our domestic animals have 'points,' ie. forms and proportions of parts, and likewise certain other properties, which are differently estimable."[53] To discern among them, and to weigh their relative values, required "patient observation and assiduous research" on the part of the breeder who "should try," he counseled, "by actual measurement to improve his eye, on which at last most persons come to depend, and with sufficient propriety, as it becomes wonderfully correct."[54] Once so trained, using this expert judgment required the dexterity of a balancing act. As Sebright explained, "We must observe the smallest tendency to imperfection in our stock so as to be able to counteract it, before it becomes a defect; as a rope-dancer, to preserve his equilibrium, must correct the balance, before it is gone too far, and then not by such a motion, as will incline it too much to the opposite side."[55] For breeders, this meant a perpetual process of keen appraisal and careful correction. Quality remained important, of course: if you wished to possess a superior strain of cattle or sheep, "you could but breed from the best," Sebright advised.[56] But the art of breeding was not so much about "always . . . putting the best male to the best female." Rather it required the ability to both meld the best qualities of a pair and cancel out their worst. "Were I to define what is called the art of breeding," Sebright wrote, "I should say, that it consisted in the selection of males and females, intended to breed together, in reference to each other's merits and defects."[57] On this careful scrutiny hinged the whole endeavor: "The breeder's success will depend entirely," he warned, "upon the degree in which he may happen to possess this particular talent."[58]

Transposition and Transmutation

In attempting this delicate balancing act, breeders and stock-raisers also needed to distinguish between the effects of external factors—climate, feed, conditions of management—and the inherent characteristics of their animals. Increasingly, the undertakings of those bent on the improvement of British husbandry complicated this task. Picking up a breed from one place and moving it to another, and the evident changes such actions wrought on individuals and their progeny, called into question the nature of the relationship between type and place. It also contributed to the general confusion over cause and effect in the mechanisms of heredity. As Andrew Coventry, author of the well-reputed *Remarks on Live Stock and Relative Subjects* (1806), remarked, some breeders, "having discovered that certain properties were less steady when circumstances were changed, have been disposed to conclude, that all are more or less mutable ... according to the influence of the changing powers." But in "a situation where circumstances were less varied, and where of course alterations on the form and character of animals were less frequent and striking," breeders "collecting their observations" there were "led to draw an opposite conclusion," the stability of the conditions of their observations suggesting that "the appropriate qualities were innate and immutable."[59] Heredity and environment were impossibly entangled.

Transpositions at the colonial scale threw this problem into sharp relief. In the early nineteenth century the paradigmatic example of the confusion born of relocation was that of European sheep transposed to the Caribbean, where a few generations seemed to wholly alter the character of these translocated ovines. "The heat has such a strange effect upon the sheep, in the West-Indies," observed one commentator, "that hair like that of a goat, grows upon them, instead of wool."[60] This appeared to be true no matter how "fine and close their fleece might have been in their native country."[61] It was, moreover, common knowledge—"a fact that all our seafaring people who frequent warm climates are well acquainted with," according to one gentleman who published under the name "Agricola." "When they export any live sheep from Europe to supply themselves with fresh provisions," he wrote in the *Scots Magazine* in 1774, "they always find, that ... what springs upon them in warm climates is only a thin straggling coat of a particular kind of hair, hardly at all resembling wool."[62]

While general interest periodicals and agricultural journals alike were rife with such assertions, other interested parties were quick to counter that the seemingly strange action of the tropical sun was rather the old action of like producing like at work. The "notion of sheep losing their wool, and becoming hairy, after remaining a few seasons in the climate of Jamaica" was a "groundless" one, as John Lawrence asserted in no uncertain terms. The perceptible changes that overtook European sheep "doubtless" came from "intercopulation with an hairy breed" already established there, and "in course, suffered to intermix with our woolly breed."[63] In fact, such ovine miscegenation was a common occurrence. Nor was it limited to colonial conditions, given how "impossible" it was, according to another expert on Scottish sheep farming, "by any ordinary care [to prevent] any strange sheep com[ing] into a different district ... from intermixing with the native sheep at the rutting-season."[64] And as the progeny of that first illicit coupling "[ran] the same risk of being farther debased than their parents were, it must of consequence follow, that, after a few generations, they will have so far lost their distinctive marks as scarce to be distinguishable from the sheep with which they are now associated."[65] The outcome, he explained, was "invariably a mongrel breed,"[66] in which local and introduced (read: improved) breeds alike were "debased" by careless husbandry, whether in Scotland or the colonies. Whatever the causation, the implication of accounts that described transplanted European sheep introduced to colonial places "assum[ing] a covering fitted to the climate, [and] becoming hairy and rough," was that they thereby degenerated from their previous heights of cultivation.[67]

Discourse of this sort bore not only the character of four-footed creatures. It played upon a more generalized anxiety about human transposition that attended the colonial endeavor, and this gave the issue a more lasting purchase than it might otherwise have had. Beginning in the early modern period, as Europeans settled in foreign places that were home to unfamiliar plants, animals, and people, they wondered at, and worried over, the potential impact such radical dislocation would have on their own bodies. The idea that human beings were sensitive—for better or worse—to the effects of their surroundings, and that some kinds of people were better suited for particular climates than others, was a long-standing commonplace in European thought, and it shaped the thinking of all sorts, from physicians to philosophers.[68] Health and well-being were particularly vulnerable to the action of environment in the Galenic framework that defined medicine and dietetics from antiquity until the nineteenth century. Ill health signaled an imbalance between body and environment, while well-being could be restored by correcting (with the aid and expertise of a physician) this imbalance. This, combined with the long-standing idea that the nature of a people (e.g., cold, hardworking) was suited to, and in some degree defined by, their geography (e.g., Northern Europe), meant that the idea of settling distant, unknown lands could be disquieting for European colonists.

Although the valences ascribed to human physiological acclimatization were not fixed—during the early period of settlement in North America, for instance, English settlers took some solace in the idea that their bodies would eventually adapt to their new surroundings, as Joyce Chaplin has shown—as the British Empire of settlement expanded into more and more distant territory (biologically as well as geographically) over the course of the eighteenth and nineteenth centuries, the vulnerability of human bodies to the influences of their environment became increasingly disquieting.[69] Such anxiety was an undercurrent that ran through discourse and debate like the one over the effects of tropical environments on European sheep, and it rose to the surface with refutations of the action of the tropics on animal bodies. The mongrelization of translocated sheep was a matter of interest and a source of worry precisely because it potentially prefigured the racial degeneration of human colonists. Breeds of livestock offered a rough analogy to races of people, and both were vulnerable to the perils of acclimatization.[70] These included the immediate risks of bodily adjustment (called "seasoning" for both animals and people) as well as the potential for typological change or adaptation on a more permanent basis. The stakes of the debate could not be higher. So when John Lawrence refuted the effects of Jamaica's sun on the wool of British sheep—"That sheep of any species, if strictly preserved from foreign intercopulation, would retain their natural characteristic wool for ever, upon the island of Jamaica... there is not the slightest reason to doubt," he chastised his credulous readers, "since such phenomenon accords with the constant tenor of our experience"[71]— in the next breath, he spoke to the perils and promise of the human colonial endeavor. "The residence of a long-haired European colony... in the heart of Afric [sic]," he wrote, in a tone typical of the nineteenth-century Briton's confidence in his innate superiority over other races, "no intergeneration with the natives taking place, would not deprive the former of the original external European character stamped upon them by nature; would not be able to transform them into woolly-headed *shangalla*."⁷²

The Extremes of the Country

What held true for people also held true for sheep, and vice versa, whether in Jamaica or Scotland, Cambridgeshire or Cornwall. And while the circulation and relocation of breeds within the British Isles never produced effects as dramatic as those of Anglo-Caribbean colonial transposition, they did significantly alter the landscape of British agriculture. Though the effects of climate and environment relative to those of heredity and selection in shaping the characteristics of a breed remained open to debate, the great enthusiasm for transposing kinds together with the rigorous interventions in heredity that marked agricultural improvement made it increasingly difficult to maintain faith in any simple one-to-one relationship between environment and type. By the early years of the nineteenth century, regional distinction had, in many cases, become confused by the efforts of improvement-minded farmers and landowners to augment their profits on the backs of exogenous breeds. In Gloucestershire, for instance, the predominant breed was "that of the Cotswolds, a type large and coarse in the wool," but Leicesters, Southdowns, Wiltshires, Somersetshires, and the Ryeland breed could all find their champions there.⁷³ In Cornwall, where "the climate and soil ... [are] particularly favourable to the production of the finest fleeces," but where wool-growing suffered from want of a local wool fair to stimulate the ingenuity of breeders, the local "true Cornish breed of sheep" had already by the close of the eighteenth century been mostly replaced by other breeds "introduced at different periods ... of the Exmoor, Dartmoor, North and South Devon, Dorset, Gloucester, and Leicester sorts."⁷⁴

No breed was more widely diffused throughout the British Isles, or more perfectly improved, according to contemporaries, than the New Leicester Longwool, the celebrated creation of Robert Bakewell. Noted for both his skill in controlling the variability of his flocks and his business acumen when it came to maintaining control over the reproductive potential of his improved breed, Bakewell "fixed" the characteristics of the New Leicester Longwool by rigorously selecting individuals displaying his desired type (animals not conforming to his ideal were sent to slaughter), and subsequently inbreeding close relations in order, in modern terms, to hone the

The New Leicester or Dishley breed of sheep. From David Low, *Breeds of the Domestic Animals of the British Islands*, Volume 2, Plate 19. Rare Book & Manuscript Library, Columbia University in the City of New York.

genotype and thereby stabilize the phenotype of his breed. Improvement in the case of the Dishley breed, as it was sometimes called after the farm at which Bakewell undertook its improvement, meant a round carcass, heavy hindquarters, light offal and bones, and animals that were quick to reach maturity.[75] The Dishley was hailed by its contemporaries as the pinnacle of perfection and the ultimate in man's ability to bend the natural to his will: "To such extreme perfection has the frame of this animal been carried," enthused Lord Somerville, a leading improver of the age, "that one is lost in admiration at the skill and good fortune of those who worked out such an alteration. It should seem, as if they had chalked out, on a wall, the form, perfect in itself, and then had given it existence."[76] So thoroughly had the New Leicester Longwool been remade in the hands of Bakewell and other improvers that it had become an entirely synthetic kind.

The Dishley thus stood in sharp contradistinction to those breeds where the effects of nature seemed to prevail more so than "the hand of man."[77] Indeed, the idea of a "native" breed was born out of, and depended upon, this contrast for its own existence. While improved breeds like the New Leicester Longwool seemed wholly artificial—its "remarkable [reproductive] precocity" was a particular mark of its "being so much artificial," as William Brown put it[78]—unimproved "native" breeds appeared stubbornly uncultivated, and in many cases even uncultivatable. So much did "certain varieties . . . seem to have been shaped chiefly by the nature of their environments," opined Brown's theoretical heirs A. H. Archer and James Sinclair at the close of the nineteenth century, that the "physical characteristics of the country" probably "contributed in perhaps a greater degree than methodological selection on the part of breeders to the production of many of our existing races."[79] As Brown himself described it, the Scottish Blackfaced breed—one of those varieties which he classified as "native" and a type hailed for its superior wool and hardy constitution—was the Leicester's polar opposite in this regard. "There is as much difference betwixt these sheep," wrote William Brown, "as there is between hothouse and hardy plants."[80] If the Dishley breed was the ultimate in man's ability to form and control nature, the other was almost wholly the product of nature itself, subject only, as Brown believed, to "selection among themselves."[81] Together, they represented "the extremes of this country."[82] "Native" breeds emerged as the dark side of improvement's moon.

Once the New Leicester Longwool was fixed as an improved breed able to produce generation after generation in conformity with its type, the breed's influence spread rapidly throughout Great Britain. Both Bakewell's methods and his breed were adopted by "enlightened" breeders throughout the United Kingdom. Some followed his lead by inbreeding their own stock in order to "fix" desired characteristics, but most crossed their preferred breed with an already improved type, usually the Dishley.[83] The desirability of the New Leicester was such that Bakewell was able to exert a monopolistic influence on the market for improved sheep-breeding. Bakewell charged dearly for the use of his stud stock and maintained tight control over their reproductive capacities. At the height of his career, other breeders could pay the exorbitant price of £300–£400 to use his rams on their own ewes for a single season, provided they agreed to castrate all male offspring, but ewes of his improved type were never available for widespread use, which meant that Bakewell alone could lay claim to completely "pure" New Leicester sheep.[84] Other eminent breeders who agreed to let their

Black-faced heath sheep. From David Low, *Breeds of the Domestic Animals of the British Islands*, Volume 2, Plate 7. Rare Book & Manuscript Library, Columbia University in the City of New York.

own stock upon similar terms (although not at such a high profit) were, for the hefty sum of £100, granted membership in the select brotherhood of the Dishley Society, and given unlimited access to Bakewell's studs. Together, these means directed the use of the Dishley's reproductive potential, and capitalized upon its "genetic template," a business development easily as revolutionary in the realm of livestock breeding as the reformulated breed itself.[85]

Notwithstanding the outrageous costs of access to the New Leicester Longwool, its influence spread far and wide. By the first decades of the nineteenth century, few breeds remained in existence without some mixture of Leicester blood. At first, even in places where doubts "as to the merits" of the new breed prevailed, such as Sherborne in Gloucestershire, "even the advocates for the old native breed allow a cross from the latter, if not carried too deep, to be an improvement."[86] As the ethos of improvement turned in the case of livestock production to efforts to increase the yields of meat, milk, and wool, the "native" breeds of the various regions of the

British Isles began to fall by the wayside. The old Cornish breed was one of the first to succumb to the influx of improved competitors. Already by 1807, a spate of such introductions to Cornwall ensured that the "pure Cornish sheep [was] now a rare animal, nor," according to Worgan, "from its properties"—including "grey faces and legs, coarse short thick necks... narrow backs, flattish sides, a fleece of coarse wool" and "mutton seldom fat"—"need the total extinction be lamented."[87]

And the influence on British stock was profound: scarcely a breed, much less a region, remained untouched by the improved Leicester. Not only the Cornish type, but "the pure breed[s]" of Gloucestershire, Norfolk, and various other localities became increasingly rare.[88] Accordingly, at just the moment local breeds were gaining recognition and wider currency, their particular traits adapted for local conditions were beginning to be stamped out in favor of more generalized adaptations to a growing national market. Crossbreeding, it was well recognized, served to sever the ties between type and locality. As Brown wrote, it was "quite possible to bring even the mountain breed"—that most different in form and habit from the Dishley—"to prefer the Leicester lands, by simple though attentive crossing and recrossing."[89] Through crossbreeding with the improved Leicester, a type formulated for the fast and effective production of meat, British breeds became increasingly homogenized as market standardization superseded the regional and even local specialization that had previously characterized British livestock.[90]

Making Meat

A national appetite for British (or sometimes for English) meat emerged in the nineteenth century concurrently with, and partly dependent upon, this standardization of local breeds. Population growth, supported by increased industrialization, higher agricultural yields, and better nutrition, supported, among other things, the emergence of a middle class in Britain beginning in the mid-eighteenth century.[91] A more comfortable income for more of the population in turn encouraged the growth of markets for staples and luxuries alike. Such appetites were discursively underpinned and furthered by the expansion and increasing specialization of the periodical press, which contributed to the formation of class-based and national tastes, both for cultural artifacts and pursuits and, increasingly, for meat. Together, these developments placed a premium on the production of British meat: increased spending power meant that mutton, beef, and lamb composed a greater proportion of more people's diets, while specialist and general readership presses increasingly emphasized this as a defining mark of Britishness.

Meat became central not only to the British diet but to a sense of nationhood and identity, a connection that was forged in no small part through the patriotic rhetoric of improvement.[92] The high status of livestock breeding in Britain was due not only to its profitability but to "John Bull's respect for his own table."[93] Those involved in this endeavor drew a direct connection between the efforts of stock breeders to perfect the meat-making capabilities of their stock and the well-being of Great Britain. A healthy population was the mark of a healthy nation, and for Britain, whose population was burgeoning at this time, a supply of adequate and wholesome food was a primary concern.[94] John Hunt held strong views about the ways in which a love of country could be expressed through agricultural undertakings. Although he was willing to concede that it was "more the business of the politician" than of a physiologist such as himself to determine "the degree of population which would be most consistent with the happiness of Great Britain," it was manifest that "the increase of population and the improvements in agriculture must of necessity be connected with each other." Thus he held that it was "the first duty of the agriculturist to make the most produce of the soil."[95] Indeed, "if patriotism [was] not an empty name," Hunt cried from the pages of a self-published pamphlet, "so long as the power, the dignity, and the prosperity of a country can be supposed to depend upon the health and happiness of a people, that the sacred character of the patriot will appear in no less splendour in the agriculturalist, who supplies the poor with wholesome food, than in the soldier who defends his country with the sword."[96]

Though rarely stated in such effusive terms, similar views on the importance of agricultural production in general, and breed improvement in particular, were widely held. John Sinclair declared this "so great an operation" that "the public alone [was] equal" to its task.[97] Improvement served not merely to line the pockets of the likes of Robert Bakewell. Rather, individual profit was tied to national gain. When a breeder was convinced "that a change of breed will suit his pasture, and be more profitable than the one he is accustomed to," Sinclair claimed, everybody benefited. The grazier "derive[d] more advantage by purchasing that sort" for fattening, and the butcher from a "carcase ... much in request with the customer he serves." The consumer benefited from a supply of superior meat "in point of taste and flavour," the currier from a "pelt or skin [that] ... answer[ed] his purpose better," and the manufacturer, for whom "the wool of the breed recommended can be worked up into better cloth, for which there must always be a greater demand, and a better price at the market."[98] Such a dense web of connection between production, industry, and consumption enabled improvers to argue forcefully for the broad social and political weight of their undertakings.

For some engaged in the work of improvement, providing food for Britain's growing population was of the utmost importance. Thomas Rudge held that "profit to the breeder, and produce to the consumer" were "the two grand objects" of improvement, and in the case of sheep, never should the improvement of wool come at the expense of "the increase of mutton." It mattered little to the farmer, he argued, whether his profits came from coarse or fine wool, or "whether his stock consists of large or small carcasses," so long as they could be made "equally ready for the market."[99] But it was "of material consequence, both to him and the public, that the greatest possible quantity of meat, with a reasonable proportion of fat, should be fed on a given quantity of land." Any other consideration, Rudge argued, "should yield to the supply of that produce which affects the support of life."[100]

The strength of these claims was such that, over time, Britishness became instantiated in the flesh and forms of these breeds themselves. Increasingly, it was what differentiated the British from other peoples. The association between national identity and beef-eating has a long and robust history, predating even the idea of Britishness itself. Steven Shapin has demonstrated the strong analogic connection between beef and Englishness by way of Galenic dietetics in the early modern period.[101] Growing focus on distinctive breeds throughout Great Britain strengthened and added nuance to this association between beef and national identity. By the mid-nineteenth century, a writer for *Chambers's Journal* attributed Britons' greater stature, strength, and "physical superiority" over the French to their "better supply of Butcher-meat,"[102] and by the close of the nineteenth century, even their foreign rivals recognized the British national talent for producing (and consuming) meat, and more than this, their ingenious ability to instantiate these traits in the very form of their domestic animals. According to one French agriculturalist, the English fondness for "roast meat" showed in the "prominent loins ... [and] small flaccid rump" of English breeds, while the "prominent and spacious" rear of the typical French breed spoke to the appetite in France for "pot-au-feu."[103] Consuming meat made Britons British. Without a steady supply of quality meat, boosters argued, "'John Bull' would soon become as watery as a turnip."[104]

The formation of a national taste for meat supported, and was supported by, the homogenization of breeds, itself achieved through the dissemination of improved varieties. By 1870, William Brown could write that "so much have pure breeds"—referring to the old, regional or local varieties that predominated at the turn of the nineteenth century—"become now intermixed, not only with each other, but with each other's crosses" that an entire volume "on the subject was warranted."[105] However, improvement had its limits, and very often the success of an introduced breed was constrained by regional climatic and environmental factors. As Brown wrote, "All improvements invariably radiate from a centre, but they do not flow equally in all directions—the soil, altitude, rainfall, and temperature, in the case of agriculture, together with man's prejudices, tending individually and in combination to turn aside or altogether dam up the regular flow."[106] The distribution of breeds, no less than other agricultural improvements, "has also been regulated by these influences."[107]

These limits held for both improved and unimproved varieties, and became increasingly evident as the influence of the New Leicester Longwool spread beyond its home county. Even for a breed like this one, that had "been made for the country, and not the country for it," certain circumstances were not wholly conducive to its thriving.[108] Some "respectable farmers" in Cornwall, for example, "still doubted ... whether sufficient advantages have been derived by their introduction," despite the relatively long use of the Dishley breed in that region. They reported that "the stock produced by the cross" did not thrive—they were "deficient in wool, particularly under the belly," they "lamb[ed] with difficulty, and [they were] bad nurses," all classic signs of what contemporaries called degeneration. Moreover, they were "too tender for the wetness of the climate" and "also liable to the foot-rot," a common complaint among breeders occupying fens, marshes, and wetlands who introduced exogenous breeds to their humid pastures.[109] As Brown put it, "even with good food, sheep cannot lay on mutton when their bed is wet and cold."[110] At the other climatic extreme, Improved Leicesters were found eminently unsuited for higher elevations. Though this breed boasted the greatest geographical range of improved breeds, its "distribution is the one with least limit of elevation"—the "alluvial plains and sandstones ... claim the whole of the Leicesters of England," their "altitude limited by 700 feet."[111] "Nothing can be more absurd, or preposterous," declared John Sinclair in allusion to the Dishley breed, "than to suppose that a fat animal, incumbered with a great quantity of wool, can ever be calculated for a hilly, and far less for a mountainous district."[112]

The effect of climate was such that even improved breeds could not wholly transcend their local origins. In some cases, this meant that "improvement" acquired a distinctly local cast. This was particularly true for Southdown sheep, a class adapted to the "peculiar habitat" of the chalky hills of the southwest of England and that underwent improvement in the early nineteenth century.[113] Despite the best efforts of their improvers, they retained "the tinge of their origin, which still adheres to them, [and] gives them a hardiness that would otherwise be remarkable."[114] Indeed, "So much do these sheep keep to the lime, that it may be safely said, were there more arable surface on the Down hills, or a much greater depth of other soil not of a chalky nature, the breed of sheep would have to be changed—probably to the Leicester."[115] So close was the relationship between the Southdown's character and its native soil that the son of its foremost improver, T. Ellman, claimed that, were the breed removed to Leicestershire, "the fine flavour of Southdown mutton may be changed in time to the coarse, tallowy meat of the Leicester or other long-woolled sheep. Nor will the flesh alone be interfered with, but the wool and every other feature will become assimilated to those of the natives of the different localities."[116]

The persistent tie between locality and type, even after the application of improved methods of breeding, also meant that the Southdown, when it did circulate, disappointed. F. Boys reported in the *Annals of Agriculture* that farmers in the neighborhood of Hillinton, Norfolk, in the east of England, some distance from the home territory of the Southdown, found the improved Southdown unsatisfactory. They were "too tender for this country, the land here being too open for them," these farmers reported, although Boys claimed that such objections were "ridiculous!," given that Ellman's own flock of 500 ewes, grazed on the Southdowns, produced 620 lambs in one season on land "as much exposed, and as open, as any lands can possibly be."[117] Despite these protestations on behalf of the breed, and although its productivity (measured by per capita meat and wool outputs) increased, the Southdown's range did not extend far beyond its home region—at least until the conditions of imperial agropastoralism made its circulation beyond Great Britain's shores to the antipodean colonies possible.[118]

The importance of compatibility between locality and type was further evident in the tendency of unimproved local breeds to languish outside of their native circumstances. Gooch saw this at work in the fens of

Southdown sheep. From David Low, *Breeds of the Domestic Animals of the British Islands*, Volume 2, Plate 14. Rare Book & Manuscript Library, Columbia University in the City of New York.

Cambridgeshire when cattle from the neighboring county of Suffolk were tried. "An opinion prevails at Islesham that the Suffolk cow will not thrive in the fens," he wrote, though the two locales were separated by less than thirty miles as the crow flies. A local farmer had "proof of it, by having purchased some from Suffolk, and having kept them with his other cows two years, during which they gradually declined." The insalubrious effect of the fens on this type was confirmed when the farmer "sold them [back] to the person of whom he bought them, and they were soon restored to their original health."[119]

This Cambridgeshire farmer's experience was a lesson that colonial breeders would learn again and again in the second half of the nineteenth century as they worked to adapt livestock bred for the various conditions of the British Isles to the dry heat of Australia, the long winters of western Canada, or the steep hillsides of New Zealand.[120] Breeds produced for one set of circumstances were not always suited to another: "the physical character of a country"—its soil, temperature, rainfall, and vegetation—had "marked influences" on the variety of kinds of livestock, "not only on those introduced from different habitats, but even on those whose constitutions have been long inured to the particular ranges where any change of climate may be brought about."[121] The natural aptitude of domesticated populations to alter in response to external conditions—be they "natural forces," deliberate selective influences, or some combination thereof[122]—was the very mutability that improvers used to reformulate their breeds for higher output in the late eighteenth century, inducing their stock to reach maturity more quickly, and to achieve greater extremes of woolliness or fleshiness. But in the colonies, adaptation to novel environments would be interpreted as a loss of control over the character of their breeds, and perhaps most worrisome, as a loss of Britishness.

CHAPTER TWO

Much Ado about Mutton

In the early years of the nineteenth century, a spirited and sometimes ugly debate broke out in the pages of the *Commercial and Agricultural Magazine*, one of Britain's foremost agricultural periodicals. This debate concerned merino sheep—a breed native to Spain that had been introduced to the British Isles during the last quarter of the eighteenth century. Then, as now, merinos were reputed almost the world over for their exceptionally fine wool—wool so fine, in fact, that when compared with that of other breeds, it "appears as silk," as the eminent agriculturist Arthur Young enthused in the late eighteenth century.[1] The "extreme beauty"[2] of this valuable article was indisputable, yet the breed's arrival in Great Britain was controversial: its form and flesh were no match for the excellence of its fleece, and even staunch proponents of the merino like the famed naturalist Joseph Banks, who was one of the breed's most elevated and most prolific supporters, had to admit that the shape of these animals left something to be desired. In 1800, he conceded that "their carcases were extremely different in shape from that mould which the fashion of the present day teaches us to prefer."[3] That fashion was for rotund, barrel-like proportions, best exemplified by the New Leicester Longwool or Dishley breed. This animal represented the pinnacle of British sheep-breeding, and was almost universally admired as "[a model] of symmetry and good form."[4] But where the Dishley had been carefully honed to embody a very particular combination of stoutness and delicacy—designed to produce the maximum quantity of fat and flesh upon the minimum of bone and offal—the merino seemed to be all waste. Their "outlandish forms" were the antithesis of the ideals of profit that governed British sheep-breeding. They were long-legged and narrow-chested, with low hindquarters and horns so "prodigious" as to give the rams "an unsightly appearance in the eyes of those who have been accustomed to hornless sheep," a category that included many "improved" British breeds.[5]

Disagreement over the merino's introduction to Great Britain pitted supporters of the Spanish interloper—most of whom hailed from the landowning classes—against champions of the New Leicester Longwool. The ensuing debate was protracted, the first clash erupting in 1802, and the last battle over the merino—in the pages of the agricultural press at least—only

Merino ram. From William Youatt, *Sheep: Their Breeds, Management, and Diseases.*

subsiding in 1812. Controversy in this particular publication was common during the decade in which the merino debate raged: contributors regularly debated the relative merits of various agricultural implements, fodder crops, and systems of management. Few other topics, however, seemed to generate quite as much prolonged vituperation as the merino controversy. As early as 1802, a proponent of the Dishley breed writing under the name "Practicus" asked, "Can a farmer ever hope to pay his rent with a flock of deformed, unthrifty, diminutive sheep, and a few tods of bastard wool?"[6] Caleb Hilliar Parry, the author of a well-reputed treatise on sheep-breeding and anatomy, declared these claims "gross misrepresentations and illogical conclusions," and called his interlocutor "flippant, declamatory, dogmatical, and expressive of the most profound ignorance."[7] Later salvos in the ongoing battle were even more hostile. When John Hunt of Loughborough sparked another round of poison pen letters to the *Agricultural Magazine* in August 1808 by calling the merino "high shouldered" and "hollow backed," the "blind zealots of the Merino cause" retaliated, insulting Hunt's views as "insipid and pointless inanities," his favored breed as "living Dishley oil barrels," and he himself in unflattering terms as "the doughty defender of Dishley blubber."[8]

The unusual duration of the controversy, and the extreme vilification of men and sheep alike that it produced, suggest that the debate cut to the quick of more issues than those merely concerning livestock breeding in Britain. Because this episode played out against a geopolitical backdrop of nearly endemic warfare with Revolutionary and Napoleonic France, ongoing conversations over the relative importance of climate, heredity, and human influence in livestock breeding took on added urgency with respect to naturalizing merino sheep in Great Britain. France's aggressive expansionism on the European continent continually threatened Britain's access to Spanish merino wool during this period, making the establishment of a domestic source of fine merino wool seem even more pressing from one point of view. The very specter of prolonged war, however, only affirmed for the merino's opponents the paramount importance of British food security, which, they claimed, rested on propagating New Leicester Longwools—not merinos—as the most efficient converters of feed into fat mutton. Thus, debate over attempts to establish the merino in Britain finally came down to the question of whether to breed for British meat or Spanish wool, an inherently irreconcilable pair of aims not only because of the recognized difficulty of selecting for fleece and flesh at the same time but because they ultimately demanded that the merino conform to the cultural and economic conditions of British breeding in the first place, while resisting the effects of environment and climate on its wool in the second.

Each side of the controversy professed to have the interests of the nation at heart, but what these were, and how they were best to be defended, were debatable. As opposing sides rallied around their chosen breeds, the intersection between class affiliation and competing visions of political economy were laid bare. Elite agriculturalists—the landed gentry who dabbled (or more than dabbled) in agricultural improvement—promoted domestic fine wool production and the consequent profits of industry as the key to national stability, while on the other side, professed "practical farmers" argued that securing the sustenance of its population with domestic mutton production was of paramount importance. In their efforts to encourage the breed, merino enthusiasts tried to have it all, taking advantage of fluid understandings of heritability and environmental influence to suggest that the combined influence of selection, crossbreeding, and the climate of the British Isles could improve both the merino's form and flesh without deteriorating its wool. But these were ambitious claims, and perhaps too sanguine, at once overestimating the skill of British breeders and underestimating the degree and significance of the creolization that attended the merino's naturalization in Great Britain.

Fleece versus Flesh

Why Great Britain should need to import and establish an entirely new breed for the purpose of growing fine wool, given the widely held belief at the turn of the nineteenth century that the British Isles were uniquely suited to the growth of sheep and wool, and the prevailing certainty that the ovine products of Great Britain were "the best in the universe," puzzled some contemporaries.[9] Since the Middle Ages, the wool of native British breeds like the "Old Herefordshire" or Ryeland breed had been widely exported, and were highly renowned throughout Europe. By the seventeenth century, though, the Spanish merino had surpassed British shortwool fleeces in volume, value, and fineness. Even Britain itself had come to rely on Spanish superfine: the shawls and broadcloths manufactured in Gloucester, Wiltshire, and Somerset relied on a blend of native and merino fine wools for their value as luxury export goods.[10] Part of this reliance, ironically, was due to the same processes of improvement that had produced the merino's ovine rival, the New Leicester Longwool. As Bakewell and his acolytes solidified the ability of native British breeds to produce succulent fatty joints, mutton superseded wool in significance for perhaps the first time in the history of British sheep-breeding. More and more, as meat production took precedence over wool, attention to the carcass—its shape, its symmetry, and the rate at which sheep reached maturity—increasingly came at the expense of the quality of the wool. Breeding for wool and breeding for meat seemed, to contemporary experts, to be largely incompatible aims, and as mutton and lamb gradually assumed precedence in much of the realm of British sheep-breeding, the quality and often the quantity of wool produced declined.[11]

But if the British increasingly bred for meat at the expense of wool, their Spanish counterparts put nothing before the growth of the merino's golden fleeces. Until the third quarter of the eighteenth century, outside Iberia little was known about the sheep that produced Spain's famous wool, or their management, despite the wide availability of their wool throughout Europe.[12] As a form of live capital, sheep (like other species of livestock) have the inherent ability to reproduce themselves, imparting their characteristics, whether the capacity to produce fat mutton or fine wool, to their offspring.[13] The Spanish crown rightly recognized that its stake in the international wool market rested on maintaining a firm monopoly on the production of fine merino wool by exerting strict control over the reproductive capacities of the breed. Its monopoly, therefore, depended on the ability to contain this generative capacity of the sheep—the same "spirit of monopoly which prevail[ed]" in Bakewell's restrictive terms of letting.[14] In practical terms for Spain, this meant tight control over the export of live animals, which was governed by the strict laws of the Mesta, an arcane and secretive corporate body that oversaw all aspects of the production of sheep and wool in Spain.[15]

However, by the mid-eighteenth century, Spain's monopoly on both the wool and the animals that grew it began to falter. The persistent efforts of agricultural emissaries, spies, and diplomats began to pierce the veil of mystery that had shrouded the sheep and their management. Early reports from Iberia, like the "Account of the Sheep and Sheep-Walks of Spain" addressed to the naturalist Peter Collinson and published in 1764, circulated widely.[16] Foreign interest centered on the *transhumantes*—the seasonal journeys undertaken each year by flocks ten thousand strong. Under the watchful eye of the Mesta, and according to an "itinerary . . . marked out by immemorial custom," these megaherds perambulated from winter lowland pastures to the Alpine regions of Leon, Castille, and Arragon and back again.[17] These peregrinations were necessary, Collinson's interlocutor (and others after him) believed, to produce the "short, silky, white wool" for which the merino was known: Spain's stationary flocks, like those of Andalusia, "who never travel," had "coarse, long, hairy wool." A "life in an open air of equal temperature," he claimed, was the key to the merino's fine wool, and accounted for the elaborate efforts expended in shepherding them to and from their high-altitude summer pasture in flocks "as well regulated as the march of troops."[18] Without these seasonal migrations, "if the fine wooled sheep stayed at home in the winter, their wool would become coarse in a few generations." By the same logic, this gentleman reasoned that if Spain's stationary "coarse wooled sheep travelled from climate to climate, and lived in the free air, their wool would become fine, short, and silky in a few generations."[19] This reasoning resonated with British breeders' experience and observation of the diversity of their own kinds of stock, and with other early French accounts of merino management in Spain, this account helped to solidify the sense of connection between climate and fine wool—a persistent one that would dog the merino into the nineteenth century despite repeated attempts to disprove it.[20]

At the same time that such reports began to circulate outside of Spain, so too did the sheep themselves. As Spain's political might waned, both in

A pair of merino sheep. From David Low, *Breeds of the Domestic Animals of the British Islands*, Volume 2, Plate 12. Rare Book & Manuscript Library, Columbia University in the City of New York.

Europe and in its American colonies, merinos came to constitute valuable diplomatic capital for the Spanish crown, and were a much-desired object of political exchange. At one time, nearly all the merinos of Spain had been the property of the monarch, but "various exigencies of state," the author of the "Account" explained, had gradually "alienated by degrees the whole grand flock from the crown," dispersing them among the nobility of Spain.[21] The first merinos to move beyond the bounds of Iberia were those acquired by Sweden in 1723, "with the view of improving the wretched Swedish breeds,"[22] but after this early start virtually no live sheep left the Iberian Peninsula until the 1760s. The generous gift of 100 rams and 200 ewes bestowed on the prince of Saxony founded the largest and most esteemed flock of merinos outside of Spain, and in 1786 Louis XVI of France acquired a sizable seed stock to establish a flock at Rambouillet outside of Paris that soon attained considerable celebrity as exceptionally fine studs.[23]

Britain, too, was eager to receive its share of Spanish sheep, but was "one of the last powers who turned their attention towards this national concern," not least because confidence in the "vast superiority" of its own British woolens had made acquiring the foreign breed seem unnecessary.[24] Once Sweden and Saxony had demonstrated the utility of home-grown merino sheep, Britain chose not to await the magnanimity of the Spanish monarch, instead acquiring a small population from the Estremadura region by way of Portugal in 1787.[25] This extraction—at best of dubious legality, and at worst an act of outright smuggling—was followed only a few years later in 1791 by a royally sanctioned gift of thirty-five ewes and five rams of the Marchioness Del Campo Di Alange's Negretti flock, "the reputation of which, for purity of blood and fineness of wool, is as high as any in Spain."[26] This "treasure" ("for such," Banks assured the Board of Agriculture in 1809, "it has since proved itself to be")[27] became the foundation of George III's royal flock at Kew.

Royal Sheep

The origins of the merino's introduction to Great Britain were thus considerably elevated: this royal flock (a small collection of animals by any measure) became the primary stud stock for the breed's propagation in the British Isles. Such exalted beginnings helped ensure that the bedrock of enthusiasm for the breed would be the landed gentry. Efforts to acclimatize merino sheep in Great Britain, both practical and rhetorical, were driven by these upper echelons of rural society, and it was they who largely absorbed the cost—measured in guineas and pounds sterling as well as in the evident although debatable degeneration of the royal flock—in the early phases of the breed's acclimatization. The utmost efforts were taken to ensure the success of the breed in Britain. This meant both safeguarding its purity—under Banks's supervision, "Farmer George's" flock was carefully "guarded against all danger of the admission of impure blood"[28]—and distributing it to the most worthy stewards. During the early years of the flock's existence, the monarch magnanimously bestowed animals on those agricultural worthies willing to undertake the experiment of their cultivation for the nominal charge of only a few guineas. Over time, as "the carcasses of the sheep ... evidently improved" and their wool "rather gained than lost in value," as Banks claimed in an 1802 report on the royal flock, the fixed price of these animals was raised to six guineas for rams, and two for ewes.[29]

Beginning in 1804, in admission of their increasing value, and as a "means of placing the animals in the hands of those persons who set the highest value upon them, and [were] consequently the most likely to take proper care of them," the royal merinos were sold by auction.[30] In July of that year, in spite of "heavy and almost incessant rain" and rather alarming defects among the sheep for sale, the first royal auction of merino sheep, held at Kew, was well attended, and the commerce brisk and profitable.[31] Bidding was opened by John Macarthur, a pioneer of Australian settlement who attended the auction to procure stock for the recently claimed colony of New South Wales. In the first transaction of the day, he expended more than £6 on a single ram, despite the fact that it was, in the polite terms of the *Agricultural Magazine*, "labouring under a temporary privation of sight."[32] Healthier rams fetched as much as thirty-eight guineas. Would-be breeders were undeterred by such apparent signs of degeneration among the king's animals. High prices and willingness to overlook the stock's defects were signs of enthusiasm for the breed. One sheep described as "at present blind" still fetched more than twenty guineas, while another suffering from foot rot—a common ailment produced by wet conditions, and to which merinos were particularly susceptible—made £12.[33] The popularity of the animals was further signaled by the haste of newly minted merino owners to spirit home their purchases. As a writer for the *Agricultural Magazine* reported in the following month's issue, one gentleman, having failed to arrange prior conveyance appropriate for an ovine cargo, rode off with his newly purchased sheep as a passenger in his chaise, "such was the eagerness of [these] buyers to bear off their lots."[34]

Improbable as it was, this scene repeated itself at the following year's sale, where, once more despite the apparent shortcomings of the breed in general (even its fiercest promoters in Britain acknowledged that merino sheep were "very far from handsome in their shape"),[35] and the king's flock in particular (high mortality from disease among the stock sold in 1804 "had been hinted at" by that year's purchasers),[36] commerce in 1805 only increased. In the opening sale of the day, a shearling ram "of the worst appearance of the whole" sold for more than £22. Handsomer animals followed, reaching prices as high as sixty-four guineas in one case.[37] As new owners loaded their purchases into carts "with an enthusiasm of the most laudable kind," one especially eager buyer was seen "helping a ram into a carriage!" and, as he was "a man of fashion," the "scene of business presented a picture of the greatest hilarity."[38]

These instances speak not only to a willingness to overlook the possibly negative effects of British climate and conditions, or to the growing passion for merino sheep in Britain in the early years of the nineteenth century. They reveal the degree to which this enthusiasm was a freak of the upper classes.[39] Merino breeding was an elevated art. The breed's proponents hailed from the highest orders. Even its humbler champions were landed farmers influential in important breed societies of the day. Benjamin Thompson, Caleb Hilliar Parry, George Tollet, and Nehemiah Bartley, for example, all had close ties to the exclusive Bath and West of England Society. Among the breed's more lofty enthusiasts, John Southey Somerville (the fifteenth Lord Somerville) possessed, in addition to his title, the means to undertake his own merino-buying expedition to Iberia. Following "the example of his Sovereign," Somerville sailed to Portugal in 1798 "for the sole purpose of selecting by his own judgment, from the best flocks in Spain, such sheep as joined in the greatest degree the merit of a good carcase, to the superiority in wool."[40] This costly undertaking resulted in "a flock of the first quality" and the approbation of his peers.[41] Joseph Banks applauded Somerville's initiative as an act worthy of "the highest commendation."[42] Indeed, any who undertook to experiment with merino sheep—"all," as Banks put it, "who honour the Fleece"—were, in the eyes of the breed's supporters, patriots of the highest order.[43]

Patriots, For and Against

Honoring the fleece rather than the flesh of these animals signaled a particular stance on the political economy of the nation. Each side of the merino debate claimed to be acting according to national interest, although they differed in their interpretations of what that meant in terms of ovine economy. For proponents of the merino sheep, it was in Britain's best interest to ensure a favorable balance of trade by securing the continued production of luxury articles for export. They couched their arguments in terms that linked the domestic production of fine wool to the patriotic defense of British industry and crucially, in light of Britain's ongoing conflict with France, to independence from foreign trade. Worry over economic dependence on foreign supplies had been growing toward the end of the eighteenth century. Modern estimates suggest that British demand for Spanish wool had grown by a factor of sixteen over the course of the eighteenth century,[44] but access to this article fluctuated during the Continental wars. Imports of Spanish wool ranged between 2.1 and 4.1 million

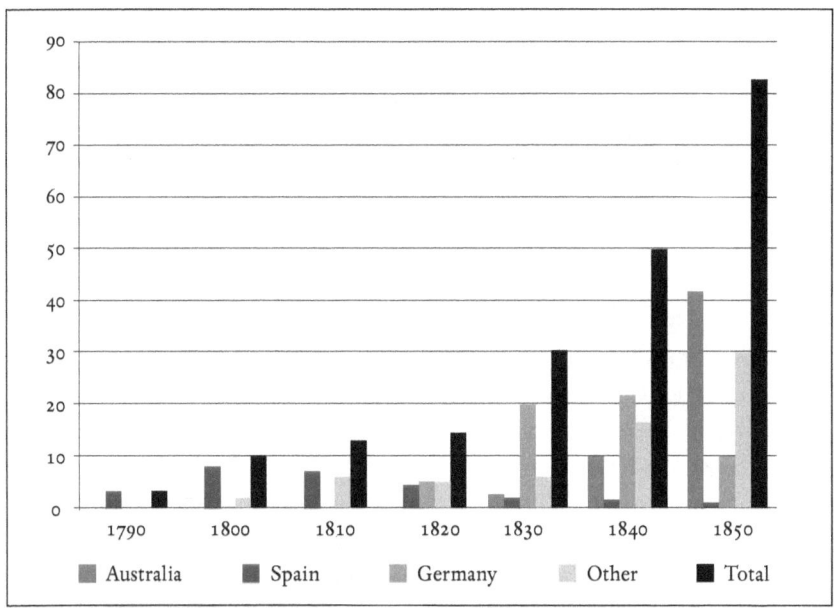

British wool imports in millions of pounds, 1790–1870.
Values are estimates, and are compiled from Mitchell, *Abstract of Statistics*, 191–93, and Carter, *His Majesty's Spanish Flock*, Appendix, fig. 8.

pounds in the late 1790s, but rallied to an average of 6.5 million between 1800 and 1806. In 1807, Britain imported an astonishing 10.5 million pounds before that figure plummeted to only 2.1 in 1808. Thereafter, volume continued to vary but never sank below 4.6 million until the close of the war.[45]

These erratic swings were a cause for major concern. Somerville gave voice to this growing unease when he warned the Board of Agriculture as early as 1799 that "the political situation of Spain may be such, as to shut out, or at least materially increase the present difficulty of importing her wool into this country; in which case, it is a matter of the utmost national importance, that the fine woollen trade of Great Britain should suffer nothing in reputation."[46] Such concerns seemed all the more pressing after 1807, when France both extended its naval blockade against Britain to neutral ships and invaded Spain.[47] To hear the breed's proponents, it seemed that Napoleonic France's sole aim in over-running Europe was to gain a monopoly over the trade in fine wool at the expense of Great Britain. Thomas George Bucke, secretary of the Merino Society (est. 1811) from 1812 to 1813, believed France's nefarious maneuverings were undertaken with the "object to make that country the emporium of superfine wools" at the expense of British industry, while Banks urged his compatriots to "resist the baneful machinations of our persevering and implacable foe" by actively encouraging the merino in Britain.[48] One of the first acts of the Merino Society after Banks convened it in 1811 was to translate a report on the propagation of the merino from the Minister of the Interior to Emperor Napoleon. The contents of this report, originally published in *Le Moniteur*, detailing plans for the establishment of ram depots and a breeding schedule to bring the population of merinos in France up to eighteen million, only confirmed such fears of Britain's "subtle and inveterate enemy."[49]

While the acquisition of merino sheep was not the foremost object in the belligerence of France, nor a monopoly over merino wool the primary aim of its blockade or of the more comprehensive Continental System designed to exclude Britain entirely from the commerce of Europe, its military machinations and expanding power did have a perceptible and important impact on how, and particularly on *where*, merino sheep were raised. Most significantly, it eroded the long-standing embargo on the export of live sheep from Iberia. By 1812, it seemed Somerville's dire prediction had come to pass: France—Britain's "implacable foe"—controlled the lion's share of European merinos, and prominent agriculturalists and economists feared that Britain's dependence on the importation of Spanish merino wool would reduce the nation to a "tributary to France for a supply of that article."[50]

The plight of the Spanish merino was indeed severe; with Iberia a major theater in the ongoing conflict between France and Britain, "contending armies . . . traversed the ancient walks of these animals, marking as their prey, and destroying for their food, every flock which they found upon their march."[51] By 1812, Spanish flocks had been drastically reduced: an estimated three-fourths were "already destroyed, and the remainder daily diminishing by rapine and neglect."[52] Not all Spanish merinos in the path of the French army were devoured or destroyed, though. Considerable numbers were rather spared "the rapacity of the French," and French occupiers found themselves free to disperse the remnants of Spain's massive merino flocks as they saw fit. The agents of French, German, and American interests fell over this ovine war booty, while Britain, as the sworn enemy of the French, suffered exclusion from the buying frenzy. Not only was Britain's own wool supply threatened, but, it seemed, her enemies were siphoning off all the valuable merinos from the peninsula. As Benjamin Thompson lamented, "a considerable portion have found their way into the vast tract of European territory under the controul of our inveterate enemy," while "a further number have been conveyed across the ocean to America, and other distant regions."⁵³ Despite a recent royal "gift" from the Spanish crown in 1809 of 2,000 sheep from among the famed Paular flock, and the fact that Great Britain had managed to spirit away as many as 10,000 merinos by way of Portugal during the frenzy, the lack of reliability in what Britain could expect to import continued to weigh heavily.⁵⁴

Members of the Merino Society thus viewed the task of establishing the merino in Great Britain with the utmost gravity and a sense of national consequence. The Society brought together noblemen and other agricultural worthies on explicitly nationalistic terms, as "a body of Britons combined in association for a patriotic purpose."⁵⁵ For these zealous improvers, patriotism began at home. Somerville took the lead in 1799 when he vowed "as an individual, bound in a particular manner to support the agricultural produce of my own country... never again to wear superfine cloth, or kerseymere, any part of which shall be of foreign growth."⁵⁶ Lest critics accuse the Society's well-heeled dignitaries of supporting the production of a mere luxury item for their own comfort at the expense of the availability of "animal food," which "an increasing population imperiously calls for," Banks and the other members, "the rank and number... [of whom were] commensurate with the great importance of the object," continually stressed the "great national as well as individual advantage" that would derive from their activity.⁵⁷

Not all patriotic agriculturalists understood their national sentiment in the same terms. Those who opposed naturalizing the merino in Britain also claimed to be acting in the best interest of the nation, but for them, patriotism meant breeding British and eating British, as well as wearing British.⁵⁸ John Hunt, the "Leviathan of Loughborough," led the charge against the Spanish breed. As a patriot of the "old-fashioned kind," he held that the "Leicestershire breed of sheep [was] a subject of national importance," and moreover that "truly patriotic views" meant a dedication first and foremost to feeding Britain's growing population.⁵⁹ No breed was better suited to this task than the New Leicester Longwool, its advocates asserted. A correspondent to the *Agricultural Magazine* writing under the name "Pastorius" rallied to Hunt's call between 1804 and 1806. He argued that the value of the Dishley breed was its ability to produce "a *much* greater quantity of mutton... on proportionably *less* food" than any other breed, making it the most efficient means of increasing the food available to a growing population without recourse to additional acres.⁶⁰ Moreover, Dishley mutton was of such extreme fatness that the laborers who constituted its "principal consumers" obtained "a much greater quantity of food from a pound of

Leicester than from an equal weight of small mutton" in the form of broths and drippings as well as flesh.[61]

Demography, not the balance of trade, lent urgency to this side of the debate. Fredrik Albritton Jonsson has shown how certain Scottish Enlightenment thinkers in the eighteenth century saw the imperatives of improvement (read: enclosure) in conflict with the need to maintain a healthy rural population in the Highlands—a particularly critical concern during the Seven Years' War when Scottish Highlanders helped fill the ranks of Great Britain's army and navy.[62] The merino debate echoed these earlier concerns, but in this case the contest between improvers and populationists moved indoors to sit at the table. Great Britain's population was growing in the early nineteenth century, and as John Hunt argued, "it is on our increasing population that we must depend for our national protection and support, and without a proper supply of animal food it would be impossible that our present state of population should be maintained."[63] Because the "pitmen, keelmen, and coal-heavers at Newcastle, Shields, and Sunderland, consume[d]," according to Pastorius, "a much greater quantity of mutton, individually, than any other men in the world," it was of the utmost consequence that the "extremely handsome, fat, and profitable sheep, the New Leicester," be granted "a peaceable existence."[64] But merino supporters refused to concede that their favorite breed posed a threat to the working-class food chain. Much like its wool, merino mutton was better fit for the table of a gentleman than for that of a common laborer. Thompson, perhaps exasperated by Hunt's proclamation, replied, "Let the Leicestershire still supply the labouring classes with the lard, of which so little goes a long way." The "naturalized Spaniard," meanwhile, would "furnish . . . meat, of far superior quality for the tables of the more wealthy" without any threat to Hunt's champion breed.[65]

True Spanish Wool

In fact, the project to naturalize merino sheep demanded much more than merely a peaceable existence for the Dishley. In the first place, for the effort to succeed, the animals had to retain the quality of their fleece, upon which their worth was evaluated. By and large, contemporaries understood this to be a matter of the breed resisting the influence of the climate and environs of the British Isles. At the same time, the dominant trend in British livestock breeding was to select for meat. Thus for the Spanish sheep to maintain their place on even the gentleman's table, the cultural climate demanded that they shed some of their foreign character and take on qualities of the English. This was to be accomplished through selective crossbreeding: by pairing merino rams with native ewes, and subsequently selecting from among these offspring, enthusiasts proposed to eventually establish a breed of "Anglo-merinos." By this "art of breeding," given time and "careful and judicious selection," these "Spaniards" could become more English.[66] The goals of British breeders were thus coterminous with the process of naturalization; they were to produce "a new race of sheep of their own making," one "with Spanish fleeces and English constitutions."[67]

Such a plan had the double virtue of improving the wool of native British sheep, while ameliorating the shape of Spanish breeds. But crossbreeding is always a risky proposition. Opponents worried that by these efforts, the merino would pollute British stock. As any type "invariably... propagate[s] its kind," wrote one anonymous Scottish improver in 1774, "if any one species be mixed with any other, the progeny will invariably be a mongrel breed, participating alike of the qualities of both the father and mother."[68] Sir John Saunders Sebright put an even finer point upon it. Even "if it were possible, by a cross between the new Leicestershire and Merino breeds of sheep, to produce an animal uniting the excellencies of both, that is, the carcase of the one and with the fleece of the other," an animal "so produced would be of little value to the breeder" since "a race of the same description could not be perpetuated." That is, a cross between two such different types would never breed true, and therefore "no dependance [sic] could be placed upon the produce of such animals; they would be mongrels, some like the new Leicestershire, some like the Merino, and most of them with the faults of both."[69] The attempt to improve domestic wool by crossing British breeds with Spanish sheep, they feared, would end by diminishing the excellence of British flesh and form.

Added to this, those aligned against the breed were convinced that it would degenerate in the foreign environment of the British Isles. Agriculturalists and breeders outside the Society, some prominent among the ranks of early-nineteenth-century improvers, doubted the breed's ability to withstand the harsher climate of England without sacrificing the quality of its wool. Indeed, climatic concerns and environmental unsuitability, unsurprisingly, were among the irrepressible John Hunt's primary objections to merinos. He firmly believed that "animals will best preserve their character on their native soil."[70] Following the same logic as Collinson's correspondent a generation earlier, he argued that if merinos were transposed to Leicestershire, for instance, "in a few years the nature of their offspring would become subservient to local circumstances, even if no crossing had taken place." This meant, in effect, the loss of the merino's character: "the carcase would improve, and the wool become coarser," thus negating the very justification for importing merinos, namely, the superior quality of their wool.[71] "If fine wool be the object," Hunt claimed, "it will not be sufficient that we go to Spain for Merino sheep; for if the character is to be preserved, it will also be necessary to bring the climate, soil, and pasturage with them."[72]

Conceptions of breed identity that privileged the effects of climate above all else increasingly came under challenge. The Scottish improver quoted above observed, for instance, that "many parts of England enjoy a climate similar to that of Hereford." If climate was indeed the determining factor in a breed's character, the sheep (and their fleeces) in these places would naturally resemble the Ryeland, "yet wool of an equal quality is not to be met with there."[73] Even more to the point, the kind of metamorphosis Hunt feared, his detractors argued, was impossible in the absence of crossbreeding, deliberate or otherwise: "As to foreign animals assimilating with the breed of the country into which they may chance to be introduced, without intercopulation or crossing," proclaimed a gentleman whose pseudonym (Cultivator Middlesexiensis) betrayed his metropolitan place of residence, "it is a gross deception, and has been repeatedly so proved by a long chain of facts."[74] Indeed, it was precisely such unsanctioned "intercopulation" that gave rise to the popular misapprehension of the role of climate, in the view of those who argued for the paramountcy of inheritance. As the anonymous Scotsman noted, "When a small number of any strange sheep comes into a different district, where there is a breed differing from themselves in any respect, . . . it is impossible, by any ordinary care, to keep them from intermixing with the native sheep." Subsequent generations "necessarily approach[ed] . . . the nature of the sheep with which they are intermixed; [thus] . . . it must of consequence follow, that, after a few generations, they will have so far lost their distinctive marks as scarce to be distinguishable from the sheep with which they are now associated."[75]

It is easy to see how such an outcome could be attributed to climate instead of inheritance. Before the age of steam transport sheep, like many other kinds of domesticated animals, had "little chance to be carried far from home."[76] The operation of heredity was thus apparently coterminous with the influence of local climate and environment, making it even more difficult to distinguish between the two. By the early nineteenth century, though, experience and observation over the course of generations (ovine, if not yet human) could be marshaled against this interpretation. As John Sebright wrote definitively in 1809, "It has been ascertained that neither the sheep nor the wool sustain any injury from the change of climate or pasture; and the absurd prejudice, that Merino wool could be grown only in Spain, is fortunately eradicated."[77] But he may have been overstating the case. Climatic determinism was not so easily put to bed, and continued refutations of the influence of climate and environment suggest that objections on these grounds continued to plague merino enthusiasts.

Ideas about climate operated not only against the merino's establishment in Great Britain. They could also be deployed in favor of the breed. The handy maxim used to make sense of the diversity of native British types—that "every soil has its stock"[78]—applied as well to the placement of merino sheep in Britain as to a topographical taxonomy of its native kinds.[79] Raised on the right type of soil, proponents of the breed argued that it would prosper rather than degenerate, and that it would do so without threat or challenge to the Dishley breed. Far from restocking the rich pastures of Leicestershire, breeders proposed improving the yield of "forester" breeds subsisting on the margins of agricultural cultivation in Britain. Often, these places were colder and more inclement than better lands used for arable farming or more intensive forms of animal husbandry, and the thickly fleeced foreign sheep seemed particularly suited to such regions. As one Nottinghamshire breeder wrote, merinos "seem extremely well calculated for our cold hilly situation: they are enclosed in a thick almost impervious coat, muffled round the eyes, and nearly to the end of the nose, and their legs down to the feet covered with fine wool of the valuable quality."[80] Breeders in Ireland and Scotland—understood to be the agricultural as well as political margins of the United Kingdom—concurred. Here, the yolk—the natural oil secreted by sheep, which on the merino saturated the fleece up to an eighth of an inch—offered what seemed a natural defense against the cold, wet climate. "The yolkiness of the Merino fleece," suggested one subscriber to the Society residing in Ireland, "and the compactness of its surface, act like an oil-cloth for its defence against rain, and fits the animal the better to endure our wet climate."[81] Indeed, this gentleman was unable to "conceive any breed of sheep better adapted to the climate and soil of the counties of Cork and Kerry, than the Merino breed." Though of similar stature to "those [sheep held] in common amongst the country people," their "carcases" were "so much more round and compact, and their wool so much more capable of paying for their keep [that] add to this their thriftiness and docility of disposition," and he would "recommend them in preference to

A pair of Exmoor sheep, or the "Forest Breeds of England." From David Low, *Breeds of the Domestic Animals of the British Islands*, Volume 2, Plate 6. Rare Book & Manuscript Library, Columbia University in the City of New York.

larger coarse woolled sheep."[82] A Scottish correspondent concurred: "The situation on my farm," he held—despite the harsh conditions of this northerly locale—"seem[ed] to agree with the Spaniards exceedingly well."[83]

Fitting the merino into Great Britain's finely variegated geography of sheep-breeding meant that marginal types and marginal places were far more likely to be slated for takeover by the merino than the Leicester or its territories. Few proponents of the Spanish breed felt it necessary or even advisable to cross with a long-wooled, heavily built breed like the Leicester, or to stock two such different breeds on the same pasture. Benjamin Thompson contended that as a breed "adapted only to luxurious pasturage," he would "let [Leicesters] there revel." The merino, on the other hand, could profitably improve marginal pastures at the expense only of the hillside breeds—smaller, hardier short-wooled sheep—not the Dishley. He advocated "the banishment of those unprofitable short-woolled animals now occupying our wolds and forests, in favour of Merino, yielding as much and as good mutton, with twice as much wool, ten times as valuable."[84] Hardly a voice was raised in defense of these "race(s) of horned mountaineers," unlike the chorus that rallied for the Dishley, that national treasure on four hooves, presumably because their breeders were less wealthy and well-connected than the elite merino advocates.[85]

The salience of climatic influence had to do with the perceived value of English-grown merino wool: for this experiment in acclimatization to work, the fleece of these transplanted Spanish sheep would have to retain its fineness as well as its value. Without these two characteristics, it was merely an inferior mutton-maker in the land of fat sheep. This was why proponents of the breed like George Bucke were so invested in the claim that the "absurd prejudice, that Merino wool could be grown only in Spain," was "fortunately eradicated," and objections to the merino based on climatic arguments had been "practically and completely set at rest for ever."[86] According to Benjamin Thompson, the first secretary of the Society, experience "having so clearly established the practicability of growing, in this kingdom, Wools equal to the article which we have been in the habit of procuring from Spain and other countries . . . it would be a waste of words to dwell on it."[87] As his successor, Bucke agreed. "English grown Merino wool," he asserted, "is proved equal to the superfine manufacture of broad cloth." Perhaps the best evidence that "Merinos stand our climate equally well with our native sheep" rested in the opinion of the lower orders, known to elite agriculturalists primarily for their conservatism and rude ignorance. According to Bucke, "Even the common farmer, those lumps of prejudice and antipathy, seem hankering after the Merino sheep. They cannot stand against doubling quantity and price of wool in a single cross."[88] Wealthy gentlemen were known to be open to agricultural experimentation; that many among the landed classes had embraced the merino was heartening, but that it was able to win over even "the most illiterate of farmers" seemed proof indeed of its capacity to thrive in England's foreign climate.[89]

But both fineness and value were contestable qualities, and in these regards the act and the outcomes of transplantation were more ambivalent than breed enthusiasts had anticipated, as those who would sell their "English grown Merino wool" discovered.[90] Compared with the high price of merino sheep in Britain, their wool offered a very poor return on investment. The inability to command what merino growers deemed a fair price— one on par with imported Spanish merino wool—plagued the Merino Society during its brief existence, and its cause provoked much speculation. One correspondent with the *Agricultural Magazine* noted, for example, that the prices in 1809 of Spanish merino wool from the region of Seville ranged from ten to fifteen shillings per pound, while the price of English merino wool sat, in the same year, closer to eight shillings, four pence.[91] The price of half-bred Anglo-merinos was even lower, despite almost unanimous testimony that even one cross improved the quality of English short-woolled fleeces. "Surely, then," this correspondent wrote, "more might be obtained for wool of the first cross than 4s to 4s 6d per lb."[92] The perceived prejudice among English wool staplers against merino wool produced in Britain constituted one of the most serious obstacles faced by merino enthusiasts, for if they could not convince wool buyers that merino wool grown in Britain was as valuable as merino wool from Spain, they would not be able to convince British breeders to discard their Southdowns and Ryelands in favor of the foreign breed.

And herein lay the paradox of naturalization: the merino's foreignness represented both the appeal of the breed, for it ensured the value of its wool, and the grounds for objection, for it threatened the sanctity of native British breeds. The struggle over pricing merino wool grown in Britain was about the degree to which location and environment inhered in the notion of a breed in the early nineteenth century. Whether or not merino wool did degenerate in Britain, breeders faced the belief that even if the wool were as fine as that grown in Spain, it was somehow intrinsically different, and therefore worth less on the market than "true" Spanish wool. While merino breeders held that the wool they sold really was "Spanish wool ... though grown in England," this conflict over pricing suggests, in fact, it wasn't. Despite the impassioned claim of George Webb Hall of the Society's Somersetshire committee that *"fine wool* will *ever* be as fine gold, so long as luxury shall exist, no matter where grown, so that *it be fine wool,* and brought to market in merchantable condition," market prices continually proved him (and Banks) wrong.[93] As Hall lamented, "Is there a single grower of Merino Wool in this extensive Society, or in the United Kingdom, who can report to it, that having produced fine wool, he has been able to dispose of it at a price that bears any relation to the price of Spanish wool?" This was a mere rhetorical question: among his acquaintance, there was not.[94] The superior value attached to Spanish wool suggests that its exoticism rated higher than wool produced in England even though merino wool, wherever grown, usually remained finer than the finest of native English wool. It appeared that for wool to be Spanish, it must have been grown in Spain. In removing the "Spanish breed" from its native pastures, British merino breeders lost the connection to location so crucial to its value on the market. In a way, then, John Hunt had been right when he wrote that for breeders to grow Spanish wool in Britain, they would have to bring the soil, climate, and pasturage along with the sheep, although for different reasons from those he supposed: it appeared to be a question of marketing as well as strictly one of physiology. Even the limited success of acclimatization undercut the value of merino wool. Ironically, by making the merino more native to Great Britain, the architects of its naturalization introduced the seeds of their own undoing.

A Touch of Class

While the influence of climate worked against the merino with respect to its wool, adaptation to local conditions in other ways remained a requirement of the breed's acceptance in Britain. At the same time that the wool market demanded that merino wool resist degeneration in the light of British climate and environment, the cultural predominance of meat-eating demanded that it conform to the standards of fat stock breeding of the day. This meant that the merino would have to become more British from the inside out, but it would need to do so without sacrificing the value of its wool. This was a tall order for both the sheep and the men who bred them, not least because it entailed, in the first place, overcoming the prejudice of most British breeders, which was dogged, and pertained not only to points of wool and climate but to the issue of form. "Their shape," acknowledged

Charles Henry Hunt, author of *A Practical Treatise on the Merino and Anglo-Merino Breeds of Sheep* (1809), "though what the greatest painters have chosen as models, is certainly not such as the English sheep-fanciers of the present day can admire." In contrast to the famous barrel shape of the Leicester Longwools, merinos were "in general rather high on their legs, flat sided, and narrow across the loins, and consequently defective in the hinder quarter."[95] John Hunt was, not surprisingly, considerably less generous. "If we proceed from the neck," he wrote to the *Agricultural Magazine*, "we shall find the Merino ram high shouldered, hollow backed, very deficient on the rump, long legs, carcase small in proportion to the height, with a weight of bone in all parts, sufficient to obliterate every appearance of perfection."[96] Other skeptics noted its narrow chest, a black mark against any animal "destined to be the food of man," because "no animal whose chest was narrow could easily be made fat," and the merino was "in general contracted in this part."[97]

Even staunch proponents like Somerville expressed doubt as to the merino's ability to overcome these deficiencies. For his part, if the merino failed to conform to this desired shape, the value of its wool, and, importantly, the flavor of its mutton, were such that the eye of the breeder rather than the form of the animal ought to be improved: "Supposing . . . that no great improvement in the shape should be obtained, it becomes to any man simply a question between his eye and his pocket; if he must have beauty, and that, too, of an unwieldy description, let him have it; but if he prefers profit . . . he knows where it may be found."[98] Most enthusiasts, however, retained their faith in the superlative effects of British skill and method when it came to remolding the merino. Banks believed that "in due time, with judicious management, carcases covered with superfine Spanish wool, may be brought into any shape, whatever it may be, to which the interest of the butcher, or the caprice of the breeder, may chuse to affix a particular value."[99]

Claiming that merino sheep had come to Britain in an almost unalloyed state of nature was a particularly effective way to reassure audiences on the point of its potential for improvement. Denying any skill or deliberation to the tradition of stockbreeding in Spain left room for the application of British skill. When John Hunt accused the merino of existing "in a state of uncultivated nature," he meant it as criticism, but Banks, Somerville, and the Merino Society turned it to their advantage.[100] As an "unimproved breed," an uncultivated form, the merino needed only the virtuosic eye and practiced hand of the British stockbreeder to bring it to that "extreme of perfection" to which they were accustomed.[101] Spaniards, "if they may be supposed to know what we call beauty," scoffed Caleb Hillier Parry, an esteemed member of the Bath Agricultural Society and early merino enthusiast, "have never attempted to produce it" in the form of their sheep,[102] and were, according to Hall, "at best" known to be "great slovens in all their agricultural operations."[103] Their single-minded focus on fine wool accounted for many of the breed's perceived defects. The whole system of merino husbandry in Spain was calibrated toward producing wool firstly, and meat only secondarily (if at all): it was eminently not designed to produce the kind of fat mutton so tempting to the British palate. From a British perspective the laws of the Mesta, which governed the *transhumantes*, and the long migrations themselves, were thus detrimental to the improvement of the breed. Somerville remarked that "it must be evident to every judge of stock, that a journey from the mountains of the north to the plains of the south of Spain, cannot be otherwise than productive of more injury to the frame and constitution of the animal, than of benefit to the fleece."[104] The Mesta only impeded "the Spaniards of attempting improvements, even if they had the disposition."[105]

This, and not the true nature of the merino, accounted for the "defective form of the animals originally imported" to Great Britain.[106] But much could be done to unlock the latent potential of the breed. Even its most egregious deformities could be recast as virtues. "There is an excellence peculiar to Merino sheep and their crosses, which has hitherto been little noticed," George Bucke claimed, and this was their narrow chests, or rather (from another perspective) their relatively heavy hindquarters. According to the very Bakewellian logic of concentrating growth in the most profitable cuts of meat, he argued that because "their hind quarters are heavier than their fore quarters, ... the greater weight of mutton" was concentrated "in the more profitable joints."[107] More generally, epicures asserted that together with proper husbandry, the superlative pastures of Britain would elevate the inherent potential of their flesh. "I cannot suppose," stated Somerville, "that the flesh of sheep of the Spanish breed, the grain of which is as fine as any we are acquainted with, properly fed from the birth, and on English pasture, will not prove excellent meat."[108]

Indeed, a difference in national taste for mutton could account for many of the merino's perceived shortcomings. "Mutton in Spain," explained Sir Joseph Banks, "is not a favourite food; in truth, it is not in that country prepared for the palate as it is in this." His compatriots might have consumed beef in greater quantities than sheep meat, but British mutton was a work of art: "our lamb-fairs, our hog-fairs, our shearling-fairs, our fairs for culls, and our markets for fat sheep" were "calculated to subdivide the education of each animal, by making it pass through many hands, as works of art do in a manufacturing concern," ultimately producing an object of such high quality that if "offered for sale, and if fat and good, it seldom fails to command a price by the pound ... dearer than that of beef." High praise indeed from a nation of self-described beef-eaters, whereas "in Spain," Banks explained simply, "they have no such sheep-fairs."[109]

There was no doubt that British improvers would rise to this challenge. Since "improvement in Spain seems out of the question," explained Charles Hunt, "we must therefore look to the enterprising spirit of this country for such amelioration, either of wool or carcase, as the Merino sheep are susceptible of." Such was his faith in the abilities of his compatriots that "the knowledge and attention of English breeders cannot fail," he proclaimed, "to effect great improvements in both these points."[110] By and large, this meant crossing native ewes with merino rams in the hope of imparting some of the superiority of British form to an animal clad in Spanish wool. And while the Dishley was the recognized paragon of fat mutton in early-nineteenth-century Britain, few endorsed a cross between such dissimilar types. Rather, most proponents advocated avoiding such "mountebank doctrines of crossing dissimilar breeds, whom nature in its infinite wisdom had set a sunder."[111] Pairing like with like by selectively breeding merino rams with native ewes of the smaller breeds of mountain sheep seemed to provide the best opportunity for improving the carcass of the Anglo-merino without sacrificing its wool. "The effect of a Spanish ram," pronounced Somerville, "on the fleeces of a horned flock, such as the Dorset, the Welsh (a sheep of neat frame), on the Wiltshire, the Norfolks, the Dartmoor, [or] the Scotch ... will be neither more or less than a very great increase of profit on the fleece, with very little, if any, injury whatever to the form of the animal."[112] If not a cross with "those breeds of heath-croppy," then the next most suitable cross was with another shortwool breed.[113] Parry preferred the Ryeland breed of Herefordshire for this work. His own experience, he claimed, "proved from actual facts the practicability of producing in England, from a cross of Ryeland ewes with Spanish rams, and without the intervention of a single Spanish ewe, wool *equal* to the finest which is imported from Spain."[114] Putting only the most rotund specimens of merino ram to native ewes would bring Anglo-merinos closer to "the present fashionable ideas of beauty," themselves the product of "many years of attentive study,"[115] while leaving the longwools in their preexisting state of perfection.

The Ryeland breed of Herefordshire. From David Low, *Breeds of the Domestic Animals of the British Islands*, Volume 2, Plate 13. Rare Book & Manuscript Library, Columbia University in the City of New York.

For the merino lobby, this promised to increase the breeder's profit while at the same time securing Britain's supply of wool and therefore its balance of trade, that ever-important point of political economy.

Accounting for Taste

Naturally, not all who involved themselves in the debate over the merino in Great Britain were as optimistic about the breed's transformative ability. Predictably, John Hunt had his say on this point. "If we are to resign our fat mutton" in favor of the merino, he feared, "our own fat must go into the bargain, and all for the sake of covering our lean sides with a fine coat made of Spanish wool."[116] More worrisome were the doubts about the ability of the merino to put on fat that came from more elevated corners of

Much Ado about Mutton

the agricultural world. The famous Thomas William Coke of Holkham—esteemed agriculturalist, breeder, and "a person inferior to none in respectability, real patriotism, and liberal attention to the rural economy of the British Empire"—also "declared himself unfavourable to the Spanish breed." Such was his stature that his objections (confined "entirely to the carcase; for the superiority of the wool over the English fine wool cannot be doubted") were enough to temporarily shake even "the good opinion" that Lord Sheffield, a vice president of the Merino Society, "had formed of the breed."[117]

Yet enthusiasts of the breed were increasingly confident in their immediate, as well as eventual, success. The merino fattened as well as any native British breed, they claimed. The "Spanish breed has proved itself superior in point of size and fattening quality," Cultivator Middlesexiensis asserted, and Spanish mutton was "the most solid, savoury, and nutritious, of any to be found in this country."[118] Hunt, moreover, could set his mind at ease as to the fate of his own fat, one unnamed proponent argued, since the "Spanish race" was apt to fatten "in an equal degree with any of our native breeds." The "admirers of fat men and fat mutton," then, "may console themselves, that they may procure as large, as fat, and . . . as well-flavoured meat, from the descendants of this breed, as the fine Leicestershire herbage has yet produced from any breed whatever."[119]

Paeans to the excellence of merino mutton were sung with increasing vigor. Benjamin Thompson recounted dining on "the saddle of a Merino-Dishley wether" with "two gentlemen, who, from their elevated rank in life, must constantly have excellent mutton on their tables." They were nevertheless "united in their praise of this joint."[120] Elsewhere he proclaimed of a Ryeland-merino cross that "better mutton was never put upon a table."[121] Similarly, albeit with more restraint, John Wright, one of Hunt's less illustrious and therefore more restrained combatants, reported from personal experience that though "I profess myself no epicure, I dined off a saddle of [merino] mutton . . . and as far as my poor judgment went, thought it most delicious."[122]

Perhaps most important, though, when it came to proving the mutton-making abilities of the breed, were its growing triumphs in the show pen. At Somerville's show in 1812, for instance, Thomas George Bucke's flock "made a conspicuous figure, not only for a fine fleece, but the promise of great size, and nearly an English form." As one report noted, "The Merinos, indeed, appear to be improving annually in size, and assimilating more to the English shape."[123] Coke, who was well known as a proponent of

Southdowns, had "stirred up a competition between the Merinos and the South Downs," but despite having "exhibited the flower of his flock ... large, and well laden with fat," the merinos took the day. The animals that trumped his own had been "pushed to the utmost point of obesity ... giving the most decisive proofs of possessing the faculty of taking on fat, and their mutton being equal, at least in point of goodness, *the palm of victory appears due to them.*"[124] Even these merinos remained relatively diminutive, but "the superior size of the Down sheep proves merely," this observer recorded, "that they are bigger, not better than Merinos."[125]

AND YET, for all this, the merino never took hold in Britain in quite the way that its proponents had hoped. It is true that, today, it is hard to find a so-called native breed of sheep in Britain without some degree of merino present in its genotype. Essentially, this represents the long tail of precisely that mechanism of introduction and "intercopulation" described by the anonymous Scottish observer in 1774. The merino, in effect, was absorbed into the existing sheep stock of Great Britain without ever becoming established either in its pure state or in the kind of fixed crossbreed advocated by its most enthusiastic backers. Even the question of whether or not its wool lost some of its fineness in the foreign climate and conditions remained unresolved in the nineteenth century. The refusal of wool brokers to give what breeders believed was a fair price was an equivocal judgment upon the merino. It might suggest a number of things—that the wool degenerated, or that supply exceeded demand. That the merino soon became so well established in Australia, where climate and environment are more similar to those of Spain, suggests perhaps it did degenerate in Britain. The first 100,000 pounds of Australian wool reached Great Britain in 1818. By 1826 that volume had risen to 1.1 million pounds, by 1830 it had doubled, as it did again every five years until 1845—reaching 24.2 million pounds.[126] By 1870, combined imports from the Australian colonies and New Zealand constituted twice the volume of wool imported to Britain from all other sources. Regardless of whether or not Australia provided a solution for the physiological problem of the merino in Britain, it almost certainly provided a solution to the cultural and symbolic aspects of this controversy. For in Australia, Britain found it could grow the vast amounts of "Spanish" wool it needed, without the impediment of hostile enemies like the French, and without threat to the sanctity and integrity of its own "native" breeds.

CHAPTER THREE

The First Breed of Cattle

The county of Herefordshire is situated on the border of Wales, with Shropshire to the north, Gloucestershire to the south, and the midland counties to the east. William Marshall, the author of the *Rural Economy of Gloucestershire* (1789), described it as "a sweetly-broken country," through which the river Wye and all its "various branchlets" wend, flanked by fertile valleys and "meadow banks ... steep enough to give beauty to the surface, and genialness to the soil; yet not too steep for the purposes of cultivation."[1] This scenic and fertile region, "rank[ing] among the smaller counties" of England, was known for its agricultural produce.[2] Orchards of various description dotted its hillsides, and rye was grown in abundance among the valley meadows. It was also home to superior livestock: Ryeland sheep, reputed for their fine, thick wool (second only to that of the merino); and Hereford cattle. Celebrated for both their labor and their flesh, Marshall described their frames as "altogether *athletic*,"[3] and their form, "as beasts of draught, [was] nearly complete."[4] Herefords were known as the "rent-payers" of the district, habitually pulling plows for five or six years before being turned over to the graziers of Buckinghamshire and other regions adjacent to London who "finished" them for the metropolitan market.[5] Besides this, they were "kindly feeders," becoming "as fat as mud" on shorter shrift than many other breeds.[6] Though their coats originally varied in color from reddish-brown to dove grey, yellow, brindled, or mottled, by the early decades of the nineteenth century, they were "principally distinguished by their white faces," which were paired with "cherry sides" and coats of "soft glossy hair."[7] All in all, Marshall enthused, they were "the first breed of cattle in the island."[8]

In the early nineteenth century, it was customary to describe breeds of cattle as "native" to their home districts. Cornwall, Devonshire, Sussex, Norfolk, Lincolnshire—each had its own distinctive type formed by climate, environment, regional economy, and by "the power of local prejudice" as well.[9] "A person who has travelled through the different breeding counties," wrote George Culley in the preface to *Observations on Live Stock; Containing Hints for Choosing and Improving the Best Breeds of the Most Useful Kinds of Domestic Animals* (1786), "cannot but remark [upon] the great diversity of opinion in the characteristic distinctions of excellence in domestic animals."[10] Breeds of cattle varied not only according to "the soil of different districts," but to "the fancies of the breeders" as well.[11] At a time at which mobility was powered by muscle (human or animal) rather than by steam, the combined force of regional variation and "local prejudice" worked to preserve the various breeds "in a state of greatest purity" at their localized epicenters: the greater the distance from this point of origin, the more "intermingled in every possible way" local types became.[12]

The connection between locality and type thus forged was intimate. In the case of the "native breed of Cornwall," as William Youatt, the widely published agricultural expert and veterinarian, explained, for example, the cattle were "very hardy," and appeared "calculated to endure the changeable temperature of this peninsular and unevenly-surfaced county."[13] Breeds were believed to "have their peculiarities, attributable to different causes"—including "mere local circumstances, of soil, place, feeding and breeding tactics," but also "the strong and marked impress transmitted from remote times in some original type"[14]—and without their native conditions, it was expected that regional types would lose their "character."[15] Thus the "West Highlander"—the shaggy, Scottish breed celebrated in London markets for its fine flesh—"must have his native hills,"[16] and the North Devon, too, its "native country"—a small, rich patch of Devonshire extending from "the river Taw westward, skirting along the Bristol Channel" before the breed became "more mixed, and at length comparatively lost" at the banks of the river Parrett.[17]

Herefords, no less than Devons or West Highland cattle, were understood to be "the native cattle of the county"—produced and defined by their connection to Herefordshire.[18] Even in the seventeenth century, the local climate seemed uniquely calculated to the production of "corne and cattle," being in the words of one author, "most healthful and the soyle so fertile . . . that no place in England yieldeth more or better conditioned."[19] Expressing the connection between place and type with elegance, George Garrard wrote that "the excellence of an animal" depended "in a great measure . . . upon the soil where it has been bred and the land upon which it was fattened. Without doubt, therefore, we are much indebted to the rich pasture by the Wye and the Lugside for that perfection which so eminently distinguishes the Herefordshire cattle."[20] Or, as one of the breed's foremost genealogists, Thomas Duckham, more comprehensively put it, the Herefordshire breed was "an acknowledged aboriginal race of cattle indigenous to the soil of the county from whence they take their name."[21]

A Hereford bull. From David Low, *Breeds of the Domestic Animals of the British Islands*, Volume 1, Supplemental Plate 2. Rare Book & Manuscript Library, Columbia University in the City of New York.

That Herefords were a distinctive type with their origins in this part of England seemed clear enough to most observers. Thirty years before Duckham made his remarks, William Youatt, too, had described them as "evidently an aboriginal breed."[22] Yet declarations like these were more than mere statements of geographical fact. Rather, they were claims to a particular kind of status made necessary by the conditions of "improved" livestock production, and in particular, by the intense rivalry between the Hereford and Shorthorn breeds of cattle. The "case of Herefords v. Shorthorns" was perhaps the most intense bovine rivalry of the nineteenth century.[23] As purebred cattle redefined the standards of livestock breeding, the "improved" Shorthorn—replete with well-positioned, wealthy proponents, and recorded pedigrees—became the benchmark by which Herefords and other less refined breeds were judged.

In the absence of the kind of official pedigrees that guaranteed the Shorthorn's breeding, it was crucial for breeders of Hereford cattle to find equally convincing, alternative measures of purity. Through their rhetorical association with great antiquity, claims to aboriginality, nativeness, or indigeneity fulfilled this requirement at the turn of the nineteenth century, as did interpretations of the breed's color and markings at midcentury. These were in fact conflicting metrics, the white face of the breed becoming its signature only through obvious manipulation of the breed's genotype, thereby giving lie to earlier claims to an unchanged character and great antiquity. The institution of an official herd book for Hereford cattle in 1846 might have put the issue to rest, were it not for the illusory nature of purity itself. That the desideratum of nineteenth-century pure-breeding was itself a construct meant that such contradictions between its metrics were less worrisome than they might have been but also that the construction, establishment, and defense of purity remained a matter of concern and debate throughout the century, and beyond.

The Butcher's Breed

In the early nineteenth century the Hereford—"that beautiful, hardy and flesh-forming race of cattle"[24]—was the outcome not only of the climate, environment, and productive regime of its home county but also of the way in which it was integrated into the London meat market. Herefordshire was known as "rather a rearing than a feeding county," its "soil . . . being neither applicable for dairy or feeding purposes."[25] Its specialized livestock economy—Hereford agriculturalists having "made it their study to breed steers and oxen," according to Duckham, "which should by their superior quality and aptitude to fatten command the attention of the distant grazier"[26]—and its position within the productive economy that served Britain's largest metropolis reflected this aspect of the county. Already by the nineteenth century, that system of production had a long and sophisticated history. As early as the seventeenth century, London's hinterland stretched as far as Scotland, cattle and sheep being driven from the farthest reaches of Great Britain to satisfy the city's demand for beef and mutton.[27] Toward the turn of the nineteenth century, demand for meat (and other luxuries) intensified as greater prosperity, urbanization, and the rise of a middle class attended eighteenth-century industrial development. As the consuming public grew, and as the middle class sought outlets for its newfound affluence, meat consumption—always a rhetorically important element of the

British diet—became increasingly important in actuality as well, and producing fat stock the stated aim and cherished goal of improved breeding in the early nineteenth century.[28]

Eating meat was at the core of national identity in Great Britain. According to William Youatt, it was one of the most ancient of national traits. "The fondness for this kind of food," he wrote in *Cattle: Their Breeds, Management, and Diseases* (1834), "on account of which foreigners sometimes attempt to ridicule the Englishman, is inherited from ancestors of the remotest date."[29] Whether or not a penchant for meat was a heritable trait, Britons were indeed enthusiastically carnivorous. The antiquarian John Kersley Fowler, author of *Records of Old Times: Historical, Social, Political, Sporting and Agricultural* (1898), recorded for posterity a particularly impressive menu he had enjoyed as the guest of a prosperous tenant farmer: "Clear soup, salmon and lobster sauce, two entrées, a saddle of four-year-old wedder mutton of his own breeding and feeding, two braces of partridges, sweets made by the ladies of the household, together with Amontillado sherry and Moët's champagne; whilst after dinner ... a splendid dessert, with grapes and peaches from his own garden, with the choicest old port and Château Lafitte claret."[30] In its combination of refinements—both domestic and foreign—the menu was calculated to impress.[31]

Those who had the means to enjoy such vast quantities and varieties of "animal food," but not the advantage of dining upon meat of their own breeding, stimulated demand for fat stock. Aided by new systems of management, the use of artificial feed—including turnips, oil-cake, and grain—and the improved techniques of selection pioneered by Bakewell and his ilk, breeders and graziers at the turn of the nineteenth century brought cattle to the peak of obesity.[32] Joseph Westcar, one of the most prominent Buckinghamshire graziers and a proponent of the Hereford breed at the turn of the nineteenth century, was famous for producing astonishingly fat oxen: one of his most enormous, and memorable, triumphs tipped the scales of Smithfield market at "nearly 300 stone" in 1799.[33] (Westcar evidently valued bulk in his person as well as his animals. Being himself a portly man, he was a fixture at fat stock shows—and their ceremonial dinners—"arrayed in all his glory of size, and shape, and fat!")[34] Indeed, the improvements wrought on breeds of cattle were measured mainly in terms of the breed's propensity for, and quickness in, getting fat.[35] Breeders and graziers debated the relative merits of size—some preferred large animals, others small, but as Culley noted, whether or not "the object of extraordinary *large size* [was] ... the pursuit of the enlightened breeders . . . the more valuable property of getting *fat* at an *early age*" increasingly became the measure of success.[36]

As working oxen, Herefords initially lagged behind the more precocious Shorthorn in reaching great weight at an early age, but when it came to attaining massive size and quantities of fat flesh, they had no difficulties. Unlike such large breeds as the Sussex, which were of a "gaunt, flat, leggy and huge boned sort . . . always suggest[ing] the idea of being vast consumers,"[37] Herefords were "kindly feeders" and were "by many good judges considered to approach the nearest to that perfect state of any of the large breeds."[38] John Duncumb, author of the *General View of the Agriculture of the County of Hereford* (1805), described the "true sort" of Herefords as possessing a "large size, an athletic form, and unusual neatness."[39] More important even than their "fit[ness] for labour," they excelled "as fattening stock,"[40] being, as Juliet Clutton-Brock and Steven J. G. Hall put it in their history of British breeds of farm livestock, "ideally suited for the new trade in store cattle for fattening near London."[41]

The Hereford breed was widely admired, numbering among the ranks of "the most picturesque cattle in England."[42] Possessed of many requisite marks of bovine beauty, the male of the breed (whether an ox or a bull) was large in size and boasted "a mellow hide, well covered with soft glossy hair," and "a moderate short head and wide forehead, from which the horns . . . spring in a straight line."[43] The female, in comparison, was relatively small, "extremely delicate, and very feminine in its character,"[44] superlative specimens exhibiting a "pleasant and cheerful countenance."[45] Sexual dimorphism in Hereford cattle was of "a more extraordinary disproportion . . . than is to be found in any other of the superior breeds," but small size was no obstacle to the production of large oxen: Youatt reported that Hereford cows were "not unfrequently [*sic*] the mothers of oxen nearly three times their own weight."[46] Critics held the cow's small stature as a mark against the Hereford breed, but proponents argued that it was essential to the breed's superiority, and other experts concurred. "Experience seem[ed] to have fully proved," wrote Duncumb, that small cows produced the best oxen: "when the cow is large and masculine in its character, and heavily loaded with flesh the ox will be coarse and brawny, and consequently unkind and tedious in the process of fattening."[47]

Single-minded focus on Hereford oxen meant that "little attention [had] been paid to the cow." As long as she "possess[ed] the qualifications that long experience has proved to be necessary to ensure success with her

A Hereford cow descended from the "Tomkins blood." From David Low, *Breeds of the Domestic Animals of the British Islands*, Volume 1, Plate 17. Rare Book & Manuscript Library, Columbia University in the City of New York.

progeny," breeders paid little selective heed to the females of the breed.[48] But even the diminutive and neglected Hereford cow was "capable of extending herself universally in a short space of time, when fattening."[49] So remarkable was the Hereford's "extraordinary merit as a beef-making breed" that it sent one anonymous enthusiast into raptures: "Look at its frame! The frame is that of the butcher, great in width and depth of the fore-quarters. Look, also at its flesh, by hereditary disposition laid thickly upon those parts where cattle of the dairy breeds are thin and wedgy."[50] Those parts—especially the hips, loins, and back, which for beef breeds should be well-padded with flesh—were also among the most valuable cuts of meat, and the ability to concentrate fat and flesh upon them was one of the signature achievements of livestock improvers.[51]

Temperament, too, played a role in the fattening abilities of a breed. Though with the passage of time the capacity of a breed for producing beasts of labor was less often a measure of value for high-end cattle, and the Hereford's "high degree of manifold utility" was subordinated to its fattening capacity in this regard,[52] the Hereford's roots as a working breed operated in its favor. Tenant farmers, it was asserted again and again, had selected (possibly only semiconsciously) for an animal that would submit with docility to the plow. This had made them "tractable, teachable, and not given to nervousness"—all of which eased the handling of large beasts in any context but were special assets when it came to fattening.[53] Herefords, in fact, seemed to strike the perfect balance between activity and quietude. They were not in the habit of "making long ranges" like some other types (notably hill breeds like the Scottish Highland or Kyloe breed of cattle, which "could scarcely be restrained by any fence" and was known for its ability to thrive "on the coarsest of pastures"),[54] nor were they restless or "constantly in motion, but feed kindly and flesh as rapidly as feed and rest will enable them to do."[55] Neither did they fall to the other extreme—that of the pampered Shorthorn, which depended upon supplemental feed to produce great quantities of flesh, milk, and tallow, and which, moreover, according to a "pithy, but true" Australian maxim, like the improved New Leicester Longwool breed of sheep, "wants to lie down and eat all round it."[56]

Specialization for beef production came at a cost, and that was the breed's milking tendencies. As a breed "most eminent for work and fatting,"[57] and with the emphasis placed on producing fat oxen for the London market, "little attention has been paid to the cow."[58] A possible explanation for the relatively small size of Hereford cows, it also meant that "she [had] obtained the character of being a bad milker."[59] Among the wondrous improvement wrought on the Shorthorn breed, on the other hand, was the ability to produce copious quantities of milk—hence the preference of "the London cow-keepers" for the breed.[60] Cattle-breeding in nineteenth-century Britain being a partisan occupation,[61] die-hard proponents of Herefords absolutely refused to cede ground to Shorthorns (or any other breed, for that matter) on any point. Enthusiasts argued that while Hereford cows produced a smaller quantity of milk, they made more milk *fat*,[62] but even Thomas Duckham conceded that with "the production of steers to meet the demand of the graziers being the chief aim of the breeders," the "milking properties" of the cow "[had] been neglected."[63] In general, to nineteenth-century stockbreeders, selecting for "beef and milk appear[ed] to be as antagonistic as mutton and wool."[64] As Youatt put it in *The Complete Grazier*, "A breed of cattle equally adapted to the shambles, the dairy, and the plough, is indeed not to be met with, and experience teaches that these properties are inconsistent with each other."[65] Improvement-mined breeders nevertheless continued to seek milk and meat as "a combination of excellencies," but neither time nor "practice" produced proof that both milk and meat could be "combined in one breed to anything approaching perfection."[66]

Simply producing meat for the new demands of the London market in the nineteenth century—never mind producing milk as well—was fraught with challenges. In one sense, consumer preference seemed to be driven by the notoriously "fastidious taste of the epicure,"[67] but there were the needs of the lower orders to consider and provide for, and in terms of sheer volume, this latter requirement would always outweigh the former. Despite their proportional irrelevancy, the higher orders and their freakish preferences exercised a defining authority over market production, and many blamed the epicure for the blubbery trend in beef production. When it came to mutton, as we saw in chapter 2, the lean, gamey variety was deemed the most appropriate fare of the upper echelons, and fatty mutton that of the working classes, but fat beef was more variously appropriate, though a more finicky article of food.

Much like their contemporaries debating the relative merits of merino and Dishley sheep, combatants exercised their views in the pages of the agricultural press. According to a satirical letter submitted to the *Agricultural Magazine* whose author claimed to represent the views of "Frugally Disposed Housekeepers," the "folly-feeding system"[68] (that is, the use of supplemental and artificial foods) produced "overfed cattle" and "grossly deteriorat[ed] the quality of our beef."[69] At least one-third of the carcass of a well-fed beast was fat, which "no christian can eat, *or knows how to eat*," complained the critic.[70] Instead "of it being the food of man," the greasy flesh of such animals was fit only for use as industrial products: "the food of coachwheels and other machinery; or, handed to Mr. *Glimmer-light* and moulded into a dapper-looking fellow—a tender hearted, *melting* soul."[71] Speaking for the agricultural interest, and as an advocate for improved cattle rearing, T. Weston defended fat cattle and their producers against "such short sighted and ungrateful alarmists."[72] Weston assured his interlocutor that in producing enormous specimens of fat cattle, it was "not quantity, merely, and a consequent reduction of price" that motivated breeders and graziers but also a desire to raise the "quality of the beef." Exhibitors of fat cattle paid "due regard to the pleasure, as well as to the profit, of your unthankful petitioners,"[73] Weston retorted, and any aspersions to the contrary were "founded only on misconception, and tend only to towards evil."[74]

Producing fat cattle approached an art form in the early nineteenth century, in which type and method both needed careful calibration. According to Weston, the "anxious wish" of improvement-minded breeders and graziers was twofold: to ascertain "what particular breed of cattle has the strongest propensity to fatten" and "to give every encouragement to that species of cattle which shews the strongest inclination to accumulate fat on those particular parts that are in peculiar estimation in the London market."[75] The artful produce of this endeavor—"the aforesaid extraordinary fat beef"— was as finely tuned as the animal that produced it, and required special care on the part of the consumer. As a product intended for the discerning palate of the epicure, it seemed that those among the newly affluent middle orders who aspired to its consumption needed instruction on the appropriate mode of preparation. "They are not to devour it in their usual way," cautioned Weston, "but to take quality for quantity," enjoying it "not . . . by the pound, but by the ounce," and ought "always to take it fasting, for this beef of high quality disdains to intermix peaceably with common food."[76]

The public was apt to underestimate the skill required to produce this fine product. In 1801, the writer of a particularly venomous letter to the *Times*, signing himself "Agricola," chided the Smithfield Club for its willingness to reward extreme fatness of the kind Weston encouraged. "If we may judge by the decisions which have been made instead upon the like occasions at Smithfield," he wrote, "the *fattest* animals are considered the *best*"[77]—a common refrain among critics of fat cattle.[78] Responding in the *Commercial and Agricultural Magazine*, Weston again defended producers against the complaints of the consuming public, claiming that the allocation of prizes was always made upon a more complex calculation: "that peculiar form in the animal which indicates a disposition to fatten, and at the same time a delicacy in the meat which it produces, the smallness of its bones . . . and likewise the size of the beast."[79]

Just what combination of size, delicacy, and fineness of bone would "yield the greatest quantity of animal food for man, from the produce of a given quantity of land," was difficult to ascertain.[80] While Weston promoted selecting for fine frames, others believed this to be a practice "founded on a very bad principle . . . for the diminution of the bone occasions a diminution of other useful qualities."[81] A writer for the *Agricultural Magazine* addressed the issue in more moderate terms, but was nonetheless "thoroughly convinced that very small bones and sinews, which generally go together, indicated small quantities of flesh, causing light weights and bad butcher's cattle."[82] Detractors of fine frames, though, were in the minority, and the combination of delicate bones and fat flesh was a mark of nearly every "improved" breed at the turn of the century.[83]

Partisans of the Hereford proclaimed that their preferred breed epitomized these aims: according to its early chroniclers, "many who viewed this animal alive" in the early days of the breed "never saw so much beef under a hide of the size, and upon so small a proportion of bone."[84] Whether Shorthorn, Hereford, or another breed altogether, the stakes of breeding for beef were high. When it came to selection and improvement, the potential cost of any error in judgment was "a loss to all parties concerned," but "eventually [fell] heaviest on the consumer," who, for lack of better options, was forced to purchase an inferior product.[85] At a time of population growth, "increas[ing] the food supply [of] the nation" became an object of "vital importance."[86] In particular, "in a country where markets demand so large a portion of animal food," wrote John Duncumb, "the improvement of those animals which supply it, becomes an object of general and great importance."[87] Breeders and graziers thus perceived that their actions pertained to issues of national significance. "With spirited and wise selection," "Herefordshire farmers," no less than breeders of other types, had raised their object of study so high that, at the outset of the nineteenth century, "the Public [was] now on the eve of receiving great and incalculable benefits."[88] The outcome of livestock husbandry—the production of nourishment for the British populace—was never far from the minds of those who bred or raised beef cattle, or from those who debated their relative merits.

Native Purity

Producing such fat cattle relied not only on new methods of management and husbandry, but on the manipulation of the animals' hereditary profile on a hitherto unprecedented scale. As discussed in chapter 1, honing the genotype of a given population was most often achieved by intensive inbreeding, and one result was a growing regard for purity of descent. As purebred types came to the fore, purity became the overriding principle of British livestock breeding. It was no longer sufficient to simply produce handsome cattle. The ability of an animal to consistently replicate its desired traits in its offspring—to "breed true to type"—was now the most important measure of the value of an animal.[89] As time wore on, pedigrees that recorded the genealogical history of individual animals increasingly served as a guarantee of this for "improved" pure breeds like Shorthorns, but in the absence of such officially sanctioned purity—the *Herd Book of Hereford Cattle* was not commenced until 1846—less refined types like the Hereford or the Devonshire breed had to rely on alternative measures.[90] One of the first such alternative metrics was the label "native," which took on new valences beyond either the geographical or the unimproved around the turn of the nineteenth century. Through its corollary, the idea of antiquity, designating a breed as "native" to a particular place came to operate as a proxy for the kind of synthetic purity produced by improved methods.

In 1885, a commentator for the *Livestock Journal* noted that "of late years[,] many old beliefs respecting the origins of different breeds of cattle have become rudely disturbed."[91] In fact, such notions had always been subject to controversy. Despite the certainty with which Duckham declared the Hereford an "aboriginal race of cattle indigenous to the soil" of Herefordshire,[92] whether Hereford cattle were, in fact, a true breed was a matter of debate stretching back at least to the 1780s. Observers were divided as to whether it was an artificial type, amalgamated out of longer-standing "true" breeds (for example, the Devonshire breed crossed with Welsh mountain cattle) or whether the Hereford was itself a breed whose origins could be traced to an original type of British cattle.

That a seemingly coherent breed like the Hereford might be nothing more than a mix of types was of serious concern to enthusiasts in the late eighteenth and early nineteenth centuries. In his first edition of *Observations on Live Stock*, George Culley wrote, "As to the Herefordshire brown cattle they are, I am pretty clear, neither more nor less than a mixture between the Welch [sic] and a bastard race of long horns, that are every-where to be met with in Cheshire, Shropshire, &c."[93] Such aspersions were a threat "to the cause of the Herefords, as a *breed*,"[94] according to J. H. Campbell, who complained angrily to Arthur Young, editor of the *Annals of Agriculture*, that Culley made the Hereford "a strange hodge-podge of Welsh and some illegitimates, that he represents wandering about some two or three counties."[95] More than this, he was concerned that Young's notice of Culley's volume, which he "ushered in with such flattering marks of approbation, and so many very high compliments," was more than the work merited, and, Campbell wrote, "must certainly add much more weight to *his*"—that is, to Culley's—"evidence with the jury, than (with submission) it seems to me to deserve."[96]

In part, these doubts reflect the confused state of "improvement" in the early nineteenth century. Even very localized types circulated, and the alacrity with which would-be improvers crossed different breeds remained a point of concern for proponents of the pure-breeding method. "Q."—a frequent contributor to the *Agricultural Magazine*—worried that the idea that "perpetually crossing varieties" was "essential to improvement ... generally end[ed] in confusion worse confounded," and—at least as important—explained why breeders, "having a cross in their own pates," found themselves forever without any "valuable stock."[97] Doubt as to the origins of the Hereford breed also reflected a simpler confusion that arose from the practice of calling a breed after its native county. To so name a type after "the county in which they chance to have been bred" was a custom "liable to inconvenience and misconception," Q. continued, and too widely pursued "without the smallest notice or advertence to the crosses of blood which may be in them."[98] As a result, one was apt to encounter various animals "called a Devon."[99]

Confusion over county monikers certainly detracted from a breed, but the issue went deeper than mere semantics. Culley was not alone in casting doubt on the origins, and by implication, the purity, of Hereford cattle. As strictures governing the purity of improved breeds like the Shorthorn were strengthened after the 1780s, mere localization seemed an increasingly insufficient guarantee of purity. As Youatt remarked, "each county has its own mongrel breed, often difficult to be described and not to be traced."[100] For a regional breed like the Hereford, whose breeders aspired to national prominence—Weston declared in 1801, more in hope than in fact, that "this breed [was] spreading very fast, and will, in a few years, exhibit their white faces in almost every pasture in this Island"[101]—this connection between county and breed could be damaging. An unnamed essayist for the *Agricultural Magazine* declared in 1810 that the Hereford "had every appearance of being a mixed breed,"[102] and even a self-confessed fan of the type admitted that "their origin has not been well ascertained," supposed by some to have been nothing more than "a cross between the South Wales runt, and the Holderness breed of cattle."[103]

Thus it was not merely a question of *whether* Herefords were produced in Herefordshire, but *when*, and indeed, *how*—and more than just locality was at stake in the meaning of "native" in the early decades of the nineteenth century. Antiquity, too, was a significant component of the debated meanings of the descriptor as it applied to breeds of cattle. Much of this revolved around notions of where and how British types had developed, a question that puzzled specialists in the nineteenth century. While the perception that to be a regional type was to be unrefined, "mongrel bred," or

A Shorthorn bull. From David Low, *Breeds of the Domestic Animals of the British Islands*, Volume 1, Plate 19. Rare Book & Manuscript Library, Columbia University in the City of New York.

even simply "unimproved," was a most damaging association of locality and type, Britons were also proud of the diversity of bovine types (much like Great Britain's ovine diversity), reflecting as it did both the unusually varied topography of the British Isles and the ingenuity of British breeders. "There is as great a variety in our breeds of cattle as [of] sheep," a contributor to the *Livestock Journal* wrote proudly in 1875; "length of horn, colour, bulk of frame, shape, and other characteristics distinguish them."[104]

Such diversity was indeed remarkable. "The breeds of cattle, as they are now found in Great Britain," proclaimed Youatt, "are almost as various as the soil of the different districts, or the fancies of the breeders."[105] Some had long horns—of a "disproportionate and frequently unbecoming length," projecting "nearly horizontally on either side," or "curved so as to threaten to meet before the muzzle"—while others were polled.[106] Some, like the

The First Breed of Cattle 91

A Longhorn bull. From David Low, *Breeds of the Domestic Animals of the British Islands*, Volume 1, Plate 18. Rare Book & Manuscript Library, Columbia University in the City of New York.

West Highland or Kyloe breed, had long, shaggy coats; others, like the Alderney or Shorthorn, thin, nearly hairless hides.[107] Some were all black, reddish-brown, or pure white; others were mottled, brindled, spotted, or "*sheeted*"—the head, shoulders, and hindquarters "appear[ing] as if they were uncovered, while there is a sheet [of a different color] fairly and perfect thrown over the barrel."[108]

Which, out of this wonderful array, was *the* original British breed was the source of "much dispute," although no one doubted that such a thing existed, and could be identified.[109] To some, a multiplicity of origins seemed a reasonable supposition given the environmental and biological diversity of Great Britain.[110] In his *Observations on Live Stock*, Culley "venture[d] a conjecture" on the subject, supposing it probable that Longhorn cattle had been the "original ... inhabitants of the open plain country; whilst the Wild breed, or perhaps the Welch [sic] and Scotch, possessed the woody, wild, and mountainous parts of the island."¹¹¹ Such a position had the advantage of both paying homage to Robert Bakewell's "improved" Longhorn breed and occupying the middle ground of a battle "stoutly fought," as Youatt noted, between the advocates of the long-horned variety and those of the middle-horned type.¹¹² Culley's contemporary, William Marshall, evidently of a disposition less given to flattery, saw it otherwise. The Norfolk, Devonshire, and Hereford breeds—varieties of the "middle horn" type (British breeds having "been very conveniently classed according to the comparative size of the horns")¹¹³—had all "sprung from the same stock."¹¹⁴ They appeared to Marshall to be "varieties, arising from soils and management, of the native breed of this island."¹¹⁵ Like Marshall, Youatt found himself "very much disposed to adjudge the honour to the 'middle horns.'" Not "derived from a mixture" of the long- and short-horned types, Youatt was quick to note, the middle-horned variety, represented by Devons, Herefords, and Sussex cattle, were "a distinct and valuable and beautiful breed."¹¹⁶ These types alone, rather than the more exalted improved Shorthorns or Longhorns, were, so Youatt supposed, heirs to the original breed domesticated by the ancient Britons.¹¹⁷

Whether Herefords descended from the original British breed, then—and if so, in what proximity—was of consequence to its value among livestock breeders and fanciers. Though seemingly a well-established and even ancient type, Herefords could hardly be *the* original British breed of cattle—this honor was usually reserved for the "wild white cattle" found at Chillingham, Chartley, and a few other aristocratic estates.¹¹⁸ These wild cattle were, in Marshall's estimation, the parent stock of the Hereford and its allied types, and were "a race of animals, which, it is highly probable, once ranged . . . in a state of nature," much as the bison still did in the nineteenth century in "the wild regions of North America."¹¹⁹ This type seemed especially disposed to give rise to flights of fancy among nineteenth-century observers,¹²⁰ and in a particularly imaginative interpretation, Youatt theorized that during successive waves of invasions, the ancient inhabitants of Great Britain had retreated before "ferocious invaders" to "the fortresses of North Devon and Cornwall, or the more mountainous regions of Wales."¹²¹ Once there, they undertook "the strict preservation of that which principally reminded them of their native country before it had yielded to a foreign yoke"—that is, their cattle.¹²² Far-fetched as this may seem as both a rationale for the preservation of type, and as an explanation of the diffusion—and diversity—of British breeds, it suggested that by comparing

A bull of the "wild" white type. From David Low, *Breeds of the Domestic Animals of the British Islands*, Volume 1, Supplemental Plate 1. Rare Book & Manuscript Library, Columbia University in the City of New York.

breeds from diverse regions, and deducing from linguistic and archeological evidence where and how different waves of migrants had moved through the British Isles, the antiquarian with an interest in livestock types could determine which of the county breeds were the most ancient.

For Youatt, the manifest tendency of ancient Britons to retreat to the mountains, woods, and "fortresses" of Devon, Sussex, Wales, and Scotland meant, by extension, that the cattle of these few choice regions were, by accident of history and geography, closer to this supposedly original type. "Everyone who has had the opportunities of comparing the Devon cattle with the wild breed of the Chatelherault Park, or Chillingham Castle," he proclaimed, "has been struck with the great resemblance in many points, not withstanding the difference of colour, while they bear no likeness at all to the cattle of the neighbouring county."[123] In these regional strongholds, he believed, the breed of cattle had been "the same from time immemorial," while elsewhere "through every district of Britain," it had "degenerated" through intermixture, ancient and modern.[124] The Devonshire breed took the palm for primacy in Youatt's view, but Herefords, he declared, were also "evidently an aboriginal breed, and descended from the same stock as the Devons"—and, therefore, positioned upon a closely related branch of the family tree of British cattle.[125] In this way, through their connection to their native region, and their presumed antiquity, Hereford cattle could begin to assume some of the purity requisite, in Campbell's phrasing, "to their cause as a *breed*."[126]

A Token of Trueness

If the association between purity and nativeness was complicated by notions of antiquity and primacy, phenotypic diversity added another layer of complexity to the meaning, signification, and consequences of the connection between the two categories. If anything, Herefords had historically exhibited a wider a range of markings and colorations than other breeds—gray, speckled, all red, yellow, mottle-faced, and so on. And each of these types-within-a-type had its champions. Partisan loyalty was such that when Thomas Campbell Eyton published the first volume of the *Herd Book of Hereford Cattle* in 1846, he found it necessary to "disarrange the work," which "decidedly should have been alphabetically arranged," in favor of an order that gave preference to the mottle-faced variety so as to placate its influential supporters.[127] Among this variety's champions was Benjamin Tomkins, one of the most illustrious of the early Hereford breeders in the eighteenth century, while Joseph Westcar had built his reputation as the "Prince of Graziers" on the gray strain: many of his "triumphs" at the London fat cattle shows were achieved on the backs of this type.[128] Thomas Andrew Knight—the well-known botanist whose interests in breeding spanned the plant and animal kingdoms[129]—also favored the "light colour" in his own herd of cattle. Although the "Knight coat" did not "survive the test of time and fashion," his reputation as a breeder of Hereford cattle was such that the "Knight blood" could still be seen to "[flow] freely" in the excellence of frame that marked "nearly all the best Hereford herds" as late as the 1880s.[130]

By the middle of the nineteenth century these variations had come to seem unsatisfactory. The commencement of a herd book for Shorthorn cattle in 1822 put pressure on the breeders of other types to demonstrate their own cattle's purity of blood.[131] Whereas the security of pure decent that an established herd book offered enabled Shorthorn cattle to retain their

The First Breed of Cattle 95

own "beautifully varied" mottled and speckled hides, by the 1840s at the latest all Herefords were "white-faced [and] ruby-hued," uniformity of type operating as a visual measure of purity, just as antiquity did rhetorically.[132]

Produced by rigorous and unyielding selection of a dominant trait, such consistency (in form, frame, and stature as well as in color) made for an impressive display whenever Hereford cattle were gathered in numbers. For Thomas Duckham, there could be "no finer sight for the admirers of cattle" than the city of Hereford's annual fair, which took place in October. On this occasion "several thousands of steers"—with their breeders as well as the graziers who occupied "the fertile pastures of Bucks, Northampton, Kent, Essex, &c."—congregated in the ancient city.[133] "Whatever may have been their original colour and distinctive marks in days of yore," wrote Duckham, "their present uniform appearance cannot fail to impress those who attend that fair for the first time with a degree of surprise and admiration in their walk through the streets of the city, to see line after line of them all displaying a similarity of character, and at once claiming each other as one family."[134] Of course, this was more than an opportunity to display the uniformity of the breed: as the moment at which steers passed from the hands of breeders to "the principal graziers in the counties near the metropolis, and there [be] perfected for the London markets," the Michaelmas fair had practical purpose.[135] But even observers with less at stake in the fortune of the Hereford breed, like John Kersley Fowler (who dabbled as a grazier as well as an antiquarian), were struck by the sight of so many nearly identical cattle. Fowler's allegiance to the Hereford's rival did not prevent him from acknowledging that the Hereford fair was "a sight that differs from anything of its class in England."[136] Thousands of cattle could be seen throughout the streets of the city, "all of one type and colour, the latter being a deep brownish red, with clear white faces and bellies, a strip of white down the spine ... and the tip of the tail." So complete was the breed's grasp on its native county that "no appearance of a shorthorn or any other breed was in the city," he wrote, "except, perhaps, a few Devons."[137]

Subtler regularities in flesh and form—a well-set tail and full "twist"; "well-sprung" ribs; a "thick and round chine"[138]—were also important elements in the consistency of the breed, and were in theory subject to as much variation as markings and coat color. But color and markings were among the most obvious and malleable of a type's characteristics, and the Hereford's predictable "cherry-sides and white faces" thus operated as an easy shorthand for genetic uniformity in the breed.[139] Precisely when the white-faced type superseded other varieties is a matter of debate. Reference to this trait was common at the turn of the century, but other phenotypic varieties persisted—in decreasing proportion—until the 1840s, when the transition to a unified appearance was mostly complete.[140] In different versions of the Hereford's creation myth (as related by Duckham in his 1863 address), the trademark white face appears in the breed by various means: the introduction of cows of a "red-with-white-face breed from Flanders" in the seventeenth century; spontaneous (if not miraculous) generation of a bull bearing the white face from an all-brown herd; or the "probable effects produced by a commingling of blood of the different classes."[141] The latter option was the most likely but also the least satisfying, and Duckham, who wished to "confine [his] remarks . . . to facts which can be proved," primly declined to "enter further into any . . . surmises" of that nature.[142]

Exactly when and how this transition occurred continued to occupy breed historians and genealogists for generations, in spite—or perhaps because—of this lack of certainty, but the particulars of *how* the white-faced type became "universally prevalent" is of less interest than *why* it seemed significant.[143] If, as evidence suggested, the Hereford could not be *the* original British type, the white-faced Hereford could at least be the original *Hereford* type. Uniform color came to be perceived as "a token of trueness,"[144] and thus Duckham argued that the absolute uniformity "of color and marks" testified to the authenticity of the breed, "[going] far to prove it to be the original breed [of Hereford cattle], let the other classes have sprung from whatsoever accidental or other causes they may."[145]

This rhetorical use of phenotypic regularity as a demonstration of purity at mid-century complicated the connection that early commentary had drawn between nativeness and purity of descent. In some ways, it was a departure from the way in which "native" status and its contingent historicity were used to confer purity upon the Hereford breed in an earlier epoch. Because indisputable evidence demonstrated the variability of phenotype in bygone days, the new red-and-white uniform worn by the breed was evidence that "considerable alterations have been effected in the breed," and recently at that.[146] The "fixity of colour in Herefords," therefore, was proof itself that "if aboriginal, Herefords have deviated very materially from their original type."[147] Despite the ways in which it seemed to contradict theories of aboriginality, and thus purity of descent in one sense, this new metric was adopted widely and enthusiastically. Nor did the novelty of this visual cue displace the existing metric of indigeneity: both uniformity and nativeness continued to operate as gauges of the breed's lineage, sometimes in place of, and sometimes alongside, each other. The apparent cognitive dissonance this required generated little objection on the part of either enthusiasts or critics, probably because purity itself was an illusory target, and therefore any efforts to verify it—whether based on indigeneity, a white face, or even pedigree—were necessarily imperfect.

Neither the shortcomings of the measure nor the illusory nature of the goal lessened the urgency of the task. The Hereford's secondary status relative to the Shorthorn meant that its purity of blood was perpetually under question, whether explicitly or implicitly so. What looked worse, or more "mongrel-bred," than a herd of cattle in which gray and mottled animals stood next to white-faced red cattle, no matter how similar in size and form they might be? Absolute consistency of coloration, on the other hand, promised to offer visible proof of parentage (if not always of pedigree), allowing, as Duckham had described, individuals of the breed "to claim each other as one family."[148]

Synthetic Purity

The white face as the Hereford breed's signature represented a transitional moment in its history, from relatively uncouth regional variety toward modern improved breed, but as both evidence *for* purity and evidence *of* change, it was necessarily equivocal. Recorded pedigrees, modeled upon those of thoroughbred horses, published at intervals by independent enthusiasts or by societies designed to collect and manage a breed's genealogy, promised instead a more reliable assurance of purity. Though they did not—indeed, could not—deliver absolute assurance, pedigrees and the herd books in which they were collected promised their subscribers a firmer semblance of purity. Their importance rested not in the actual information they collected, but in the effect they had on the practice and perception of livestock breeding.[149] These documents listed an animal's forebears for at least three or four generations, and because a herd book was a closed loop—all individuals in a given genealogy being verified by registration in early volumes—it gave an impression of stability to what was very much a moving target. Even though the early volumes of Coates's *General Shorthorned Herd-Book* did little more than "pin down which animal was which" for Shorthorns, the breed's well-established and respected herd book accounts for much of its popularity in the nineteenth century.[150] Next to this guarantee, even the uniform red and white of the Hereford seemed unsatisfactory.

But the problem was less with the measure than with the desideratum. Perhaps the greatest irony of nineteenth-century livestock breeding was that purity was not a natural attribute inherent in a group of animals; variation was. As Charles Darwin understood it, the work of a breeder "unintentionally exposes his animals and plants to various conditions of life, and variability supervenes, which he cannot even prevent or check."[151] In pursuing a notion of purity, then, breeders were continually swimming against the tide of inherent variation. This required that they "select and preserve," and shape such variation—the bedrock of domestication, as Darwin understood it—through rigorous observation and merciless rejection of "worthless individuals."[152] Therefore "purity," in the sense in which livestock breeders used it, required the firm hand of human intervention to be produced. Such efforts brought "the will of man" to the fore,[153] and breeds so formed showed more "adaptation to his wants and pleasures" than to any natural forces at play.[154]

But the means to this end—that is, inbreeding—remained controversial. Many objections were founded in concern over what breeders described as a resultant loss of constitution. "The great obstacle to the *improvement* of domestic animals," wrote George Culley in his *Observations on Live Stock*, "seems to have arisen from a common and prevailing idea amongst breeders—that no bull should be used in the same stock more than three years" for fear that the herd would become "*too near akin*, and the produce will be *tender, diminutive*, and liable to *disorders*."[155] But others, as Culley complained, took their objections further, having "imbibed the prejudice so far as to think it *irreligious*; and if they were by chance in possession of the best breed in the island, would by no means put a male and a female together that had the same sire, or were out of the same dam."[156] Such narrow-mindedness was, in his view, a detriment to improved agriculture, and by extension, to the national good.

Whether inbreeding was an asset or detriment to the national herd, and similarly whether crossbreeding was an aid or a hindrance to agricultural improvement, remained stubbornly unresolved—not least because of the continued flexibility of "breed" as a concept. As the popularity of each method waxed and waned in proportion to the other, either could seem ascendant. In the early decades of the nineteenth century, Culley believed that the time was at hand when, fortunately "for the public," breeders "whose enlarged minds were not to be bound by vulgar prejudice" realized that such objections were "without any foundation."[157] Crossing breeds might have been appropriate for sage experimenters like Bakewell, but once established as pure and inbred, it was the responsibility of subsequent breeders to maintain this purity. The pendulum continued to swing between extremes of cross- and inbreeding, arriving back at the point of Culley's views later in the century: "It is scarcely necessary," wrote a contributor to the *Livestock Journal* in 1885, "to point out that what was practicable in the early days of an admittedly composite breed would be simply ruinous now that the race has been firmly established, and it seems evident that the adoption of crosses of the kind would quickly result in the destruction of the edifice that has been reared by a century of careful breeding."[158]

While periodic infusions of fresh blood were admittedly necessary to maintain health and vigor, improved breeds themselves were held up as evidence of the benefits of inbreeding, not least because of a prevailing belief that "the tendency of nature is ever to revert."[159] If that was so, the closer the relation between the animals that together composed a breed, the more likely a chance reversion or "throw-back" would still resemble its fellows. But if the genetic profile of an animal was unknown or various—that is, composed of crosses between different types—the tendency toward reversion introduced an unacceptable degree of uncertainty. The offspring of an animal of questionable breeding—no matter how perfect in form or pleasing to look at—could at any time and without warning revert to an inferior type contained in its history. Containing this uncertainty was, in effect, the rationale for pure-breeding. To breed from an animal of mixed or unknown parentage was to risk "produc[ing] nothing better than mongrels,"[160] according to the American stockbreeder G. T. Turner—the very thing that the extreme consanguinity to which animals were bred, by narrowing the genotype of a group of animals, was designed to forestall.

Appearance and observation alone were thus increasingly insufficient guarantees of consistency, and official genealogies, modeled upon those of thoroughbred horses, were increasingly required to verify the purity of a breed.[161] These published pedigrees conferred enormous prestige, as well as monetary value, upon individual animals and their breeders, and over time they became an end unto themselves.[162] Taken too far, the pedigree could be a pernicious force in livestock breeding. In no breed was the "abuse of pedigrees" more egregious than in the case of the Shorthorn.[163] "What has given Shorthorns their very exceptional value?" one contributor to the *Farmer's Magazine* asked sarcastically. "Not their intrinsic value (merit?) alone, but the ability of the owners to point to a long line of pedigree.... Pedigree is no doubt all very well, but a long pedigree on paper is not always a good one in fact."[164] Similarly, as another commentator complained, for Shorthorn fanciers, breeding for practical ends was "beside the question." When it came to highly bred bulls, "their blood is priceless, and they are to get bulls and heifers for sale as blood stock, for stud purposes again; [thus] the bullock"—that is, the animal destined for slaughter and human consumption—"is a very remote contingency."[165]

All too often, this meant sacrificing symmetry and constitution for pedigree. Deviation from the desired type and imperfection were the outcome of this lamentable habit, degeneration the unintended consequence of refinement. A writer for the *Livestock Journal* complained in 1875 that "even at this day, after so much has been done for their improvement," the "breeder of Shorthorns . . . finds quite a variety of formations in his females, very few of them approximating a perfect model." Defects "in constitution and formation," he continued, "cannot be remedied by the use of a male possessing similar ones, however desirable his pedigree."[166] Indeed, such "fashionably-bred animals" were "notoriously bad beasts." Having been "bred so long without proper judgment and from nearly related blood, [t]hey have become ewe-necked, weasel-waisted, leggy, and consumptive."[167] More moderate commentary, too, worried that Shorthorns had become little more than a fancy breed. "Shorthorn breeding in England," wrote a contributor to the *Farmer's Magazine*, "has fallen, for the most part, into the hands of gentlemen who have made a hobby, a 'fancy,' or 'fashion' of it; and who treat their hobby precisely on the same lines as the tulip hobby, or the antique bookbinding hobby."[168] In pursuit of the right pedigree, "their object [has become] purely fanciful; certain strains of blood are pronounced 'fashionable,' and straightaway the ideal is fixed on producing families possessing this blood in an intensified form by breeding its individuals *in terse [sic]*."[169]

Many, though, maintained the value of the breed. Shorthorns remained popular: of that, there was no doubt. Even if critics claimed that the passion for "'fancy' or fashionable strains of Shorthorn blood" had reduced them to "the veriest weeds, with nothing whatever to recommend them to the bullock breeder,"[170] they still boasted other "valuable perfections," especially when it came to producing milk. No other breed could "stand the London treatment"—that is, the close "cow-houses and hot food" that characterized milk production for the metropolis—and still give large quantities of milk better than the Shorthorn.[171] And their popularity would persist. "Times will boom for most things of any national importance, ranches will rise and fall, the dairy interest fluctuate, and live stock trade generally will go and come according to supply and demand," wrote a philosophically minded contributor to the *Livestock Journal* in an almost bittersweet tone at the height of the agricultural depression of 1885, "but no form of national trouble will ever lower the blood of Booth or Bates in the eye of John Bull."[172]

The End of Things Bovine

Despite continued appreciation for Shorthorn cattle, critics of this breed complained with some justification that its breeders had lost sight of the fact that "the end of things bovine is beef."[173] Hereford breeders, on the other hand, were in no danger of such a lapse. Unlike its more "cosmopolitan" rival, their favored breed was an eminently practical one, even if a number of its supporters were "gentlemen of the first rank."[174] Commentators in the late nineteenth century celebrated the breed's humble roots, one writer for the *Irish Farmers' Gazette* claiming that "the credit of the breed has been upheld solely through the judgment and skill of the tenant farmers, who have always been [its] principal breeders."[175] These lowly origins, though, ought not to be overstated. Many of the tenant farmers who bred Herefords, though not of the landowning class, were themselves prosperous.

Even the most illustrious Hereford herds were founded upon practical principles. For instance, the Hampton Court herd was founded in the early nineteenth century by Sir Hungerford Hoskyns "for the gratification of no fancy aims."[176] On the contrary, Hoskyns had established the herd in a much more public-minded spirit: "to breed bulls that could be supplied on reasonable terms for the use of the tenants on the estate."[177] By the 1880s, the herd had become the property of his grandson, John Hungerford Arkwright, who was also the inaugural president of the Hereford Herd Book Society in 1878.[178] This group of animals was lauded throughout the nineteenth century for the perfection of its constituent individuals, many of whose pedigrees went "to the very roots of Hereford genealogy."[179] Seventy-three bulls had been used since the herd's establishment, and to peruse a list of them was to see "a pretty complete epitome of Hereford history," according to a journalist for the *Livestock Journal*.[180] Its illustrious pedigrees notwithstanding, since the very inception of the Hampton Court Herefords, tenants had been "at liberty to send their cows for service ... by the very best bulls that the herd could produce," originally at "nominal fees," and then, under the even more public-minded grandson, "gratis."[181]

This utilitarian spirit defined the Hereford breed, and as a consequence, the shift to pedigrees as a measure of value was uneasy. Thomas Campbell Eyton, the ornithologist and naturalist who first began the *Herd Book of Hereford Cattle* (1846), initially met with strong opposition.[182] Many breeders insisted that an animal's pedigree was "on his back"—that evidence of a good frame "well covered with superior flesh ... should in itself be a sufficient guarantee."[183] By this reasoning—widely employed by Hereford men—the proof was in the pudding, and writing down the recipe put the cook's reputation at risk. As Eyton attempted to "hunt up all proved thoroughbred cattle and register them in a permanent and authorised volume,"[184] he discovered that many breeders, following Bakewell's example, preferred to "jealously [guard]" their methods "as a profound secret," fear that published pedigrees "would show too much of the system they pursued in breeding."[185]

The problem with pedigrees was not simply that they might expose selective practices that breeders preferred to keep proprietary. It was more fundamental than this. The insubstantial nature of purity itself meant that the entire edifice of the pedigree stood on shaky foundations. Even where breeders were willing to cooperate, pedigrees had to be "built up," the life history of an animal gleaned from private records or from memory.[186] The process was an "arduous undertaking," vulnerable to honest error (among other kinds) even in the best of circumstances.[187] Memory itself was a notoriously fallible faculty. Eyton cited dependence "upon the memory of breeders" as a source of error in the first volume of the *Herd Book of Hereford Cattle* (1846),[188] and twelve years later when Duckham published the third volume, it was still a problem, many animals having been included "with very short pedigrees." However, "it [did] not follow," Duckham cautioned, "that they [were] not purely bred." Rather, short pedigrees resulted from the "want of proper entries in private Herd Books," and "the fact of too much being entrusted to the memory of their breeders, at whose death their knowledge ... departed with them."[189]

Private herd books, in which a breeder recorded the births, deaths, pairings, purchases, and sales of all his animals, could mitigate the shortcomings of memory or "pocket-book memorandums."[190] John H. Arkwright was known for his meticulous record-keeping, "the whole system of private herd-book keeping at Hampton Court" being, in the words of a writer for the *Livestock Journal*, "the most elaborate and perfect I have ever come across."[191] Arkwright kept "a careful record of the bulls used in the herd each year," compiled comprehensive lists of cows, and made notes of the date of birth for every calf.[192] This material—which together supplied "authentic information of the composition of this first-class herd"—appeared in a range of documentary forms, from the slips of paper and pocket notebooks, so easily lost, to private herd books and charts drawn up in the expert hand of a clerk, and elegant private catalogues produced for circulation among acquaintances and interested parties.[193]

Such record keeping was strongly encouraged. A private herd book was less "liable to loss or destruction than memory," which could "seldom be trusted as to pedigree beyond the immediate sire and dam"; was "generally fallacious as to dates"; and "when correct, its store of knowledge is lost to posterity at death."[194] Not even private documentation as the basis for entry into the breed's herd book could be an absolutely reliable guarantee of breeding, though. Joseph Russell Bailey, a member of parliament for Hereford, a minor Welsh nobleman, and an avid breeder of Hereford cattle, served on the Herd Book Society's editing committee throughout the 1880s. Bailey astutely observed that any regulations the Herd Book Society instituted on this point were bound to be imperfect. Were such documentation required for entry to the *Herd Book*, a step the society contemplated taking at various points in the 1880s, Bailey had no doubt that "many Private Herd Books will no doubt be concocted for the occasion." Despite the way in which this possibility exposed the easy fabrication of pedigrees, and by extension the illusory nature of purity, the editing committee, according to Bailey, "must wink at that," should it occur, as it would perform the desired effect "of getting them"—that is, private herd books—"started."[195]

No wonder, then, that the work of "get[ting] the Hereford pedigrees straight"—an effort for which Bailey had "been most anxious to do all [he] could"—was so difficult.[196] The only way to "get them quite straight is that all pedigree breeders should enter all their animals,"[197] but not even the fact that this might only be accomplished by fabricated records deterred proponents of the *Herd Book* from their work. "Concocted" pedigrees, and the Hereford Herd Book Society's willingness to "wink" at them, reveals the degree to which the system of monitoring pedigrees—like the concept of purity at its core—was a hollow one. Nonetheless, the tide was against those who resisted the imposition of an official herd book, and as the century progressed, opposition waned, reflecting the changed productive and economic realities of British livestock breeding. By the 1880s, exclusion from the *Herd Book of Hereford Cattle* had become tantamount to major pecuniary punishment, as the breeder Percy Powell complained upon the rejection of his bull. The animal's only fault was a lack of documentation, it being "ever so good an animal" and "ever so well bred," and the decision to reject it on the grounds of insufficient documentation was, in his view, "a very arbitrary despotic and selfish" one, as well as a loss of the £500 he might have had from the sale of it as a pedigreed animal.[198]

Given the apparent willingness of Bailey to bend the rules when it came to "build[ing] up the pedigrees of the breed,"[199] the rejection of Powell's bull seems arbitrary and despotic, indeed. But despite off-the-record recognition of the ease with which pedigrees could be fabricated, the Herd Book Society maintained the integrity of its official publication. Bailey asserted that "no pedigree can be absolutely certain except so far as it can be traced in the *Hereford Herd Book*,"[200] but even as recorded pedigrees and herd books sprang up among the various British breeds of cattle, both the illusion of purity and the constructed nature of breeds themselves became increasingly evident. No matter how vociferous the insistence on the value of the herd book, the undeniably ephemeral nature of the aim would always undercut any absolute claim to purity. In part, this had to do with the alchemical production of breed, which was always manifestly a composite of other varieties and other factors, yet more than the sum of its parts. "The thoroughbred horse is something more than an Arab, modified by selection, soil, and climate; and that remarkable fact, the Shorthorn, is something more than an immense expansion of a local variety," wrote a regular columnist in the *Livestock Journal* in 1885. "Every breed of every kind has had (as we believe) crosses within a century, and ... our horses, cattle &c., are all mixtures. . . . They are, one and all, compounds of recent date as much as is a plum-pudding." It was evident, therefore, "that there is not really a single 'pure' breed in existence in Great Britain of man or beast, or bird," and this writer held, "It would add to our powers of advancing recognised types did we admit this truth."[201] As sensible as this position seemed, it remained a sticking point—economic, cultural, and even emotional value seeming to ride on the question of purity of descent from the nineteenth up through the twentieth century.

The First Breed of Cattle

PART II | Greater Britain

CHAPTER FOUR

Native Colonials

In 1928, G. H. Holford, author of a slim pamphlet called *The Corriedale: New Zealand's Own Breed*, celebrated the Corriedale as "the most successful new breed of the past century."[1] As a fixed, true-breeding cross between English longwool varieties like the Leicester, Lincoln, and Romney sheep, and the merinos that were the earliest imported ovine inhabitants of the Australasian colonies, colonial stockmen in New Zealand had produced the Corriedale during the last quarter of the nineteenth century by means of just the kind of rigorous and judicious selection so often celebrated by their British counterparts. Its frame was blocky: according to the breed standard of the Corriedale Sheep Society, the brisket was "deep and wide," lending the sheep "a very square appearance."[2] But by no means had its fleece been sacrificed for the sake of achieving "a rectangular block of meat,"[3] for as much as it made "a large and well-shaped leg of mutton,"[4] the Corriedale was covered with copious, high-quality wool characterized by "remarkable evenness" in "length, density, and quality."[5] In this combination of superior meat and wool, the Corriedale, hailed as "a triumph of the sheep-breeder's art,"[6] represented—at long last—that which had eluded an earlier generation of agricultural improvers: an English sheep clothed in Spanish wool.[7]

To its supporters, the great practical utility of this combination and the fine points of the breed were manifest, but in the partisan world of stockbreeding, where loyalty to type ran deep, not everyone could be relied upon to value the new variety's merits. Critics might "reasonably" ask, Holford admitted in a companion pamphlet—*The Corriedale: New Zealand's Contribution to the Sheep World* (1924)—whether there had in fact been the "need for the evolution of another breed."[8] After all, when the Corriedale was first developed in the mid-1860s, by Holford's estimation "there were close on forty distinct breeds of sheep in the British Isles alone." A "large percentage" of these had already "been tried" in the colony, and several had become "firmly established":[9] the *New Zealand Farmer*, the colony's foremost agricultural and pastoral periodical, noted with pride in 1892 that there were "at least ten distinct breeds of sheep in the colony," nine of which originated in Great Britain.[10] Why, then, bother with the Corriedale at all?

Corriedale sheep, date unknown. Original photographic prints and postcards from file print collection, Box 7. Ref: PAColl-6001-47. Alexander Turnbull Library, Wellington, New Zealand.

The most obvious answer lies in the same logic that underlay the eighteenth-century maxim that "every soil has its own stock."[11] Merino sheep suited some parts of the new colony, particularly the high elevations of the South Island, and the English breeds—primarily longwooled types bred for the rich pastures of Leicestershire, Lincolnshire, and Kent—took to the moister, more verdant lowlands. None of these, however, were bred for the native grasses and tussock of New Zealand's largely "unimproved" pastures, and so none (Holford answered his imaginary critics) was "so suited as the Corriedale to much of the sheep lands" of the colony.[12] As colonists worked strenuously to transform New Zealand's ecology with imported grass and fodder, breeders worked vigorously to mold an ovine type suited to local conditions.[13] Efforts were motivated by an understanding of the colony's climate as mostly temperate, yet still strenuous, and by the recognized demographic cost of relocation. Though the colony was, on the whole, meteorologically more pleasant than the British Isles, "in point of climate," wrote a contributor to the *New Zealand Farmer* in 1891,

"'New Zealand' is a long word,"[14] and "the only regular thing about" its weather was "its variability."[15] Combined with the extensive character of colonial husbandry, this meant that a breed had to be a rugged one to flourish in much of New Zealand—especially its colder South Island. At the same time, the perils of long-distance translocation meant that individual animals suffered in the "seasoning" process upon arrival, and that whole populations risked "degeneration" as a result of the relative genetic isolation that marked colonial Australasia. Pastoralists wanted a robust breed to match the rigor of New Zealand's climate, and one that would not need continuous reinfusions of parent bloodstock to maintain its vitality. These imperatives came to be reflected in the "true Corriedale," which gave "at once...the impression of a hardy sheep," and possessed "a distinctive character" and a "bold outlook."[16]

Neither merinos nor British classes of sheep, moreover, were ideally suited for the changing conditions of colonial production. At mid-century, these conditions had been determined by the imperial wool trade alone, for which merinos, with their quantities of fine wool, were a lucrative choice of breed. By the early 1880s, however, producing sheep for a new trade in frozen meat with Great Britain was fast becoming a productive imperative. The need to balance these two products—meat and wool—in a breed tailor-made for local climate, topography, and environment motivated efforts in experimental colonial breeding. These efforts ultimately produced a purebred type that embodied the tension of the imperial system. With two hooves planted firmly in the antipodes, the other two stretching toward the metropole, the Corriedale straddled the competing demands of colonial pastoralism: the need to adapt stock to the conditions and realities of new lands and new climes, and the imperative to suit consumer tastes at home, which in the late nineteenth century meant producing British meat from British breeds.[17]

But the Corriedale performed a rhetorical function as well. Perhaps its most significant claim was to being "entirely a New Zealand production"[18]—a breed "native" to the small archipelago angled across the roaring forties, dividing the Tasman Sea from the South Pacific. As bold as the breed's outlook was the claim to nativeness its boosters made on the Corriedale's behalf. Even if only in retrospect, to call the hybrid type "New Zealand's own" was consequential, as any such claim to nativeness was in the context of settler colonialism. It cast the breed as a proxy for legitimizing European settlers in a way that would have been difficult, and not necessarily desirable—though no less imperative—for human colonists to make at the time.

The Sheepman's Paradise

Ovis aires was among the defining forces of colonialism in Australia and New Zealand. Although the pastoral transformation of Australasia was not a foregone conclusion, sheep were, as Sarah Franklin writes, "essential vectors" of colonization, representing the colonial vanguard of European settlement in the antipodes, where their physical presence helping settlers lay claim to land wrested from indigenous peoples.[19] The settlement of New South Wales in particular, Franklin argues, "was largely a settlement by sheep."[20] Wherever domesticated animals accompanied European colonists, they caused ecological and social disturbance, but in this corner of the empire, whose geological history diverged from the rest of the world's before the evolution of mammals, their introduction was particularly disruptive.[21] Plants and other animal species that had coevolved in the absence of mammalian types in both Australia and New Zealand were vulnerable to the grazing habits of ovines, as well as to predation from other introduced species. Throughout Australasia, surface soils were susceptible to compaction under the hooves of introduced ungulates, nowhere more so than on the island continent. Here, the combined effects of millions of such creatures disrupted plant life cycles and distribution, endemic marsupial patterns, and aboriginal lifeways keyed to already fragile and irregular meteorological patterns.[22]

None of these consequences, though, were obvious at the outset of British colonization.[23] The first sheep arrived in Australia in 1788 on the colony of New South Wales's "first fleet," as part of a complement of old world biota orchestrated to transform this distant continent into a repository of penal labor and capital growth for the British Empire. Although all of these earliest ovine colonists succumbed to the rigors of the climate, the lengthy voyage, or both, within a decade New South Wales was home to a thriving population of some 35,000 sheep. The "prodigious facility with which sheep of the Merino blood"—and other kinds—"propagate[d] in suitable climates" remained noteworthy several generations later.[24]

The circumstances of geography and the character of early colonization, though, were markedly different in and around the archipelago of New Zealand, where early European presence was defined by whaling and sealing rather than by organized plantation capitalism. Introduced domesticates in the eighteenth century—sheep included—were therefore intended as what Juliet Clutton-Brock describes as "walking larders," the descendants of seed stock deposited on offshore islands by passing seafarers.[25] From these protopastoral origins, in the first decades of the nineteenth century, sheep populations expanded to serve missionary activities on the North Island and increasingly permanent coastal settlements on the South Island, but the flocks—of Australian origin—remained in the mere hundreds until the early 1840s when efforts to establish sheep farming at scale took hold, first in the North Island, and then in the South Island.[26]

The temporal lag between Australia and New Zealand in this process meant that the wool industry in New Zealand was slower to get off the ground: while Australia provided its first million pounds of wool to Great Britain in 1826, New Zealand's return was a mere pittance.[27] By the 1870s, though, New Zealand's sheep industry had become a critical component of the colonial economy. Armed opposition to British rule on the North Island (the so-called Maori Wars stretched from the Treaty of Waitangi in 1840 until the early 1860s) and the discovery of gold in Otago and on the west coast of the South Island in the 1860s contributed to the growth of the pastoral industry on the South Island.

As settlement progressed, colonial producers benefited from the ready market in Great Britain to which they had access on favorable terms. "The ties, financial as well as domestic, which bind the colonies to this country," wrote the managers of the London staplers' firm of Windler, Bowes and Co., were assumed to work symbiotically: they "naturally" drew colonial "produce to her ports, and London has become the depôt for the distribution of the wool grown in Australasia and the Cape of Good Hope."[28] In the early decades of antipodean colonialism, while the global market for wool was strong, times were good for colonial sheep husbandry, and pastoralists on both sides of the Tasman Sea viewed the "natural increase" of their flocks favorably. Australia, especially, quickly became established as Great Britain's emporium for fine wool. From a modest preliminary export to Great Britain of only 167 pounds of wool in 1810, by 1879, the yearly tally had grown to nearly 300 million pounds.[29] According to the authors of a handbook on Australian sheep husbandry, the success of this colonial export had elevated Australia from "a comparatively unknown place" to "the greatest wool-supplier of the world" within the span of only a few decades.[30]

This advance in the colonies' fortunes was thanks to their combined flocks, composed almost exclusively of merino sheep. The small handfuls of merino sheep brought from the Cape Colony by Captain John McArthur in 1796 were the seed stock for these antipodean multitudes, to which were periodically added imported bloodstock from Britain and Saxony, and later the United States.[31] So wonderfully suited to Australia's sere climate and expansive terrain did these sheep seem, and the land to them (one observer called it "the finest wool-growing climate known" in 1870),[32] that the original lack of ovine species on the island continent came as a "curious fact" to later commentators,[33] and for decades the colonies' vast flocks of merinos grazed equally vast sweeps of land, their hooves reshaping native ecosystems as their golden fleeces produced both metropolitan and colonial prosperity.[34]

New Zealand's suitability for *O. aires* was no less a "curious fact" than Australia's; only the types that thrived in New Zealand's more varied regional climates—which were on the whole colder, wetter, and more temperate than in Australia—were, likewise, more various. As a relatively extremophile type, the merino preferred conditions that few other breeds could withstand—great heat or severe cold—but flourished in little in between. After its introduction in 1843, it therefore inhabited only the cold, high altitudes of New Zealand's craggy Southern Alps. The lowlands and river valleys of much of New Zealand were too wet for the erstwhile Spanish sheep, who suffered foot rot in such damp conditions, so British longwools like the Leicester, Lincoln, and Romney Marsh breeds were adopted to graze the islands' moister regions. The clip from these flocks did not return such high prices as those of merino flocks, but to some extent quantity could make up for a relative lack of quality, and during the heyday of the wool trade they made otherwise unprofitable lands lucrative. Later, when the transoceanic shipment of frozen meat became possible, these breeds became valuable in their own right.

Although Australia had a near monopoly on the production of merino wool, New Zealand's reputation as a "Sheepman's Paradise" exceeded that of its nearest neighbor.[35] Colonists boasted of "our magnificent climate and grand pastures,"[36] although the pastures, if not yet the climate, were a construct, as Eric Pawson and Tom Brooking point out, forced by fire, plow, and an "empire of grass" out of New Zealand's native ecology.[37] Nevertheless, in the 1880s, so amenable to sheep did the colony's climate and environment seem that individual animals who escaped the clutches of husbandry and went feral seemed to improve, rather than to degenerate, as feral domesticates were generally presumed to do in new world conditions. For example, a flock of seven "Dorsetshire white-faced ewes and rams" that had "suddenly disappeared" from the Kauweranga Valley east of the Firth of Thames was discovered seven years later, having increased in both size and number. Fifteen sheep of "enormous proportions," equal in size to "two or three ordinary sheep," and covered in "enormous coat[s] of wool," were sighted "quite accidentally" by a prospector named John Liddel, and the flock's proprietor, happy to be reunited with his wayward sheep, reported to the *New Zealand Country Journal* that he believed the original seven had multiplied to "some sixty or 100." The size of the stray sheep and their multitudinous increase offered proof positive of "the value of the New Zealand climate."[38]

Given the apparent suitability of this part of the globe for sheep husbandry, the distance between the Australasian colonies and the markets in Great Britain and Europe, and the relatively tiny size of colonial human populations, emphasizing wool production in Australia and New Zealand initially made perfect sense. Not only was wool renewable, it was lightweight and never went bad; consequently it was easy to store and cheap to ship. Moreover, its production was labor intensive only at specific times of year, namely, at lambing season and during shearing, and so was well suited to the low population density of the Australasian colonies, which was as defining a characteristic of the antipodes in the nineteenth century as were their massive flocks. In comparison to Great Britain, according to *Chambers's Journal of Popular Literature, Science and Art*, in 1883, "Australia, New Zealand, and Tasmania are exactly three times as well supplied as we are with wool and mutton."[39]

Indeed, in the ratio of sheep to people—as in geographical location—Australasia was "almost the antipodes of the British Isles."[40] At the turn of the twentieth century, Great Britain was home to almost 41.5 million people and 31.5 million sheep, an apparently healthy ovine population but one that had in fact declined 7.5 percent since the 1860s.[41] Thanks to its empire, as a writer for the *New Zealand Farmer* boasted, Britain could claim nearly two-fifths of the world's population of sheep,[42] and as it was also home, at that time, to more than 360 million human subjects, if one was inclined to exclude the populous Indian subcontinent from the imperial family (as this writer was), then the ratio of sheep to people within the empire's scattered holdings around the globe stood at an impressive three sheep for every one person.[43] The ratio of "woolly people"[44] to the regular kind was even more remarkable in the Australasian colonies, where it had exceeded more than twenty-five sheep per person since the 1870s.[45] In New Zealand alone at the outset of the 1880s, almost thirteen million sheep grazed the South and North Islands combined, a staggering number in comparison to the fewer than half a million human inhabitants.[46]

Though other extractive resources contributed to the balance sheet of the antipodean colonies (especially timber, in the case of New Zealand, and increasingly, mineral wealth in both places), colonial economies in this part of the empire were heavily reliant on the wool grown from these ovine multitudes. As such, they were vulnerable to fluctuations in the international market value. While prices were good, the flocks of New Zealand and Australia swelled. But with prices for wool in flux over the course of the nineteenth century, "that very increase ... [became] a source of embarrassment" to pastoralists and the colony alike.[47] Particularly in the 1860s when the value of a fleece dropped precipitously, and with so much capital tied up in the bodies of their stock, sheep farmers saw their profits literally consumed in feeding them.

The region's impressive ratio of sheep to people now began to feel like a burden to colonial sheepmen. Once an animal was past its wool-bearing and reproductive prime, there was little local outlet for its terminal products. Especially in New Zealand, which lacked any sizable conurbations, local populations could consume only so much mutton. During early discussion of the possibilities for engagement in the new refrigerated meat trade, Matthew Holmes, a prominent colonist, and member of parliament,[48] calculated (very generously, as it turned out) that two million sheep "would be more than enough for local consumption" on an annual basis.[49] With its 490,000-odd people to feed in the whole colony in 1881, this number represented a whopping 408 pounds of sheep meat per capita, per year for colonial consumption—a mighty sum indeed.[50] By comparison, people in Great Britain, who had one of the highest rates of meat consumption in the world, ate approximately 110 pounds of all classes of butcher's meat (beef and pork included) per capita, per year.[51]

The population of the colony, in fact, could absorb nowhere near this volume, which stood even higher before the advent of the frozen export trade. After their wool was harvested, the waste of sheep was therefore significant. Writing in 1918, William Soltau Davidson, the former manager of the New Zealand and Australia Land Company (NZALC), an Edinburgh-based firm influential in the early development of the pastoral industry in New Zealand, recalled that the surplus stock on the company's estates was so numerous that they "erected yards at the edges of cliffs, into which some thousands of these old sheep were driven, so that they might be knocked on the head and thrown over the precipice as a waste product."[52] Extreme measures like this may have been out of the ordinary, but very little could be done with sheep past their wool-bearing and breeding prime.

Relief measures for the problem of surplus sheep were as unsatisfactory in Australia as they were in New Zealand. The carcasses of fine-wooled Australian merinos, whose fleece, "when woven by English looms into wondrous fabrics... may help to dress a Duchess," as a colonial writer for *All the Year Round* speculated, were "doomed to go to pot."[53] The hindquarters of these animals were sold locally (Australia benefited from the markets provided by several large cities, namely, Melbourne and Sydney); the rest of the carcass pressed into boilers and cooked down to its fat.[54] Tallow thus extracted found local and export markets for use as candles, as soap, and as an industrial lubricant, and the remaining matter was used as fertilizer. This process made effective use of all parts of the animal, but "boiling down" excess sheep offered only slim profits.[55] Mutton could also be tinned and salted, but, as *All the Year Round*'s Australian observer admitted, "meat sold in tins" was "not popular," as "folk like to see what they are eating."[56] This was especially so in Great Britain, where consumers not only liked to regard their viands, they also "preferred the real thing"[57] over what was "rather stringy stuff, with all the virtue boiled out."[58] Consequently, the market for preserved meat remained "necessarily limited," mostly confined to provisioning the shipping industry, civilian and naval.[59] Such schemes to preserve meat or make use of carcasses for purposes other than alimentation provided a "stop-gap" for colonial pastoralists in both Australia and New Zealand, alleviating to a small extent the "want of an outlet" for colonial flocks, but these measures were, in Davidson's words, an "unprofitable relief."[60]

A Meat Famine in the Metropole

While the antipodes appeared to be "over-run with cattle and sheep,"[61] without a local outlet in sight, Great Britain faced an alarming paucity of fresh meat. By the late 1860s, experts and the public alike feared that Britain's "vast and ever-increasing population,"[62] spurred by a second wave of industrial development, would outstrip productive capacity. Although the livestock industry was in fact doing relatively well compared with the rest of the agricultural sector, domestic supply had begun to fall short of demand.[63] The nation's population was growing, incomes were rising, and consumers were increasingly willing and able to spend money on meat.[64] At the same time, the repeal of the Corn Laws in 1846, signaling the end of agricultural protectionism and the introduction of free trade, drove up the price of meat.[65] Rising prices were only exacerbated by adverse weather and zoonotic disease outbreaks in the mid-1860s that contributed to "demographic malaise" in Great Britain's livestock population and high stock mortality.[66]

The confluence of stagnating production and growing demand was described, with restraint, by the *Times* as "a matter of serious national concern."[67] A more sensational account declared the nation's meat deficit to be "something alarming" in 1868, "being, for Great Britain, over 3,500,000,000 pounds annually," or 156,250 tons short of "the quantity deemed necessary by physiologists."[68] That such anxiety over a pending "meat famine" coincided with an actual rise in average meat consumption in Britain (from 90 pounds per capita in the decade 1861–70, to 110 pounds per capita the following decade) gave greater credence to such fears.[69] As the pace of home production slowed relative to population growth and demand, the difference was made up by foreign meat, imported live or as chilled dead meat from Europe and America.[70] This meant that while one out of every twelve people was fed by foreign meat in 1867, by 1887 one in every four relied on imports to supply their tables with joints of beef and mutton.[71]

But shifting toward reliance on foreign meat was hardly less disquieting than the threat of undersupply. It carried material risk: reliance on potentially hostile trade partners, and in the case of live imports from outside the island kingdom, the danger of exposing domestic herds to contagious diseases like foot-and-mouth disease.[72] Worry over undersupply of meat in Britain in the 1860s and 1870s went beyond the metabolic. Consuming flesh was one of the primary ways in which Britons distinguished themselves from neighbors, rivals, and competitors. *Chambers's Journal* reported in 1877, "We should confidently say that no people on the face of the earth are such connoisseurs in good beef and mutton as the English, down even to the humblest classes,"[73] and it was not uncommon for Britons to attribute their greater stature, strength, and "physical superiority" over their perennial rivals the French to their "better supply of butcher-meat."[74]

And while Britons mostly proclaimed themselves a nation of beef-eaters, mutton held second place in its heart. In point of fact, despite John Bull's "grossly carnivorous" and "almost exclusively beef-eating" reputation, "the leg of mutton share[d] almost equally with the sirloin the honours of forming the piece de resistance of the dinner-table of the average Britisher," according to the *New Zealand Farmer*.[75] Even if Britons consumed two times as much beef as mutton, as the fellow writing for this publication estimated they did, the national fondness for sheep meat was a more distinguishing characteristic even than beef-eating. Other nations, after all, also consumed beef in quantity (although perhaps not as much quantity as the British), but "no other race of people ... makes the flesh of sheep so important a part of their daily food."[76]

Such a strong association between nationality and meat-eating produced a continuous demand that the growing gap between productive capacities and consumptive demands be supplied by good British meat—a demand very much in tension with Britain's growing appetite for meat. The political tenor of such commentary meant that from the British perspective, not all foreign sources were created equal. Even if they hadn't grazed the rich green pastures of Britain, as products of cultural, political, and economic offshoots of Great Britain, colonial imports were bound to be better than more alien sources. Colonial producers recognized the metabolic and cultural importance of meat to their metropolitan brethren, and the opportunity it presented, if only they could get their surplus meat, "not dismembered, and in tin cans—but whole, and in prime condition,"[77] to what promised to be "the greatest frozen meat market in the world."[78] As *All the Year Round*'s Australian commentator put it, "We had far rather [our flocks] should feed our brothers in the grand old fatherland" than be boiled down to tallow, or ground up into fertilizer. "You want mutton and beef. We want to send it to you. How can this be done?"[79]

To Bridge the Hemispheres

The simple answer was by means of new refrigeration technology, capable of arresting the processes of decay and holding meat "in what one may call a state of suspended animation" for the duration of a trans-hemispheric voyage.[80] Various means to achieve this end were the subject of experimentation throughout the 1860s and 1870s, from dry air compression engines to those that relied on ammonia absorption. Early efforts to engineer artificial cold, and to apply it to the preservation and shipment of meat—a textbook example of a perishable article—had mixed results.[81] The earliest shipments of chilled beef from the United States, which reached London's Smithfield market in 1874, simply used coal-powered fans aimed at blocks of ice, or salt and ice in the process of liquefaction—"one of the most ancient methods employed for artificial cooling"[82]—to cool the ships' holds and their cargos.[83] This system, however lucrative for the Americans, was untenable for a trade between the antipodes and Britain, as it not only relied on more space between each suspended side of meat than could be profitably afforded over the much longer journey from Australasia but also left shipments too vulnerable to the heat of the tropics.

When refrigerating engines capable of reducing the temperature in a ship's hold low enough to maintain the carcasses of sheep frozen solid were developed,[84] however, "a new vista opened before the colonies."[85] Commercial refrigeration in the form of cold storehouses and refrigerated railcars was coming into use by the 1860s, but refrigerated shipping began only in 1877, when the first "completely successful" shipment of frozen meat from Buenos Aires arrived in France.[86] Two shipments from Australia confirmed the viability of the new trade shortly thereafter: the *SS Strathleven*, which left Sydney in December 1879, reaching London in February 1880, and the *SS Protos* from Melbourne in the same year. These early shipments from Australia were hailed as "successful experiment[s]" in New Zealand, demonstrating the viability of the new trade and offering "proof that before long, Australia and New Zealand would have ships trading to and fro . . . and this must result in great benefit to these Colonies."[87]

New Zealand was quick to make good on the opportunity suggested by Australia's early triumphs. The advent of refrigeration meant, in the words of New Zealand's delegate to the Fourth International Congress of Refrigeration (1924), "the very breath of life to us."[88] The first cargo of frozen sheep meat from New Zealand arrived in London in May 1882, after ninety-eight days at sea. The pioneering carcasses that made up the inaugural cargo were dead but not frozen when they were boarded onto the *SS Dunedin*: no apparatus of any kind for freezing meat then existed in the colony.[89] Thus when a crankshaft on the ship's refrigerating engine broke during the loading process in December 1881—a "serious mishap," according to the *Otago Daily Times*[90]—the ship's voyage had to be delayed, and the residents of the city of Dunedin (roughly ten miles from Port Chalmers, where the *Dunedin* was moored) became the first to dine upon the colony's frozen mutton.[91]

Upon replacement of the crankshaft, the carcasses of 4,311 sheep, 598 lambs, and twenty-two pigs were loaded into the insulated hold of the *Dunedin* in January 1882, and the ship set sail from Port Chalmers. The NZALC, which backed this endeavor, was pleased to find the meat, upon arrival in London, had retained its "nutritive value" and was almost universally edible.[92] Soon after this initial success, the extension of steam shipping to New Zealand greatly sped up the voyage between the antipodean colony and Great Britain, and the trade in frozen sheep meat grew swiftly over subsequent decades. The quantities of meat involved swelled to nearly two million carcasses per year in 1890, and more than five million by 1910.[93] By the same year, more than 800 vessels had been outfitted "and adapted for the transport of frozen meat and other comestibles," 189 of which served the trade between the antipodes and Great Britain,[94] so that by the early

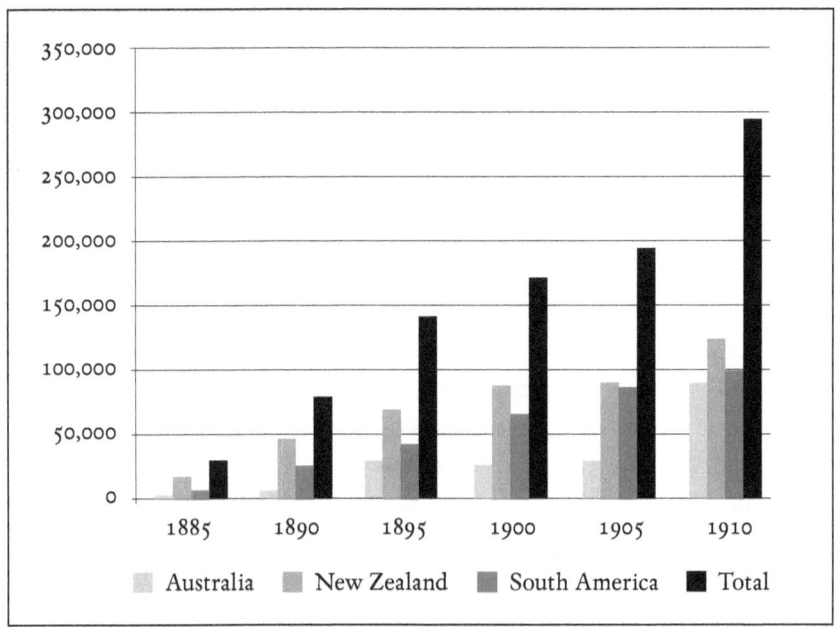

Tons of frozen sheep meat imported to Great Britain, 1885–1910.
Adapted from Critchell and Raymond, *History of the Frozen Meat Trade*, 422. 1890 and 1895 totals both include imports from the Falkland Islands (331 tons and 632 tons, respectively).

decades of the twentieth century, the trans-hemispheric traffic in frozen mutton and lamb was a commonplace for both consumers in Britain and producers in New Zealand.

As the trade itself was established, freezing works—factories for the slaughter, partial butchery (carcasses were skinned, bled, and beheaded before shipment), and freezing of sheep and lambs—sprang up throughout the antipodes, while refrigerated warehouses, or cold stores, began to populate the docks of Liverpool, London, Bristol, and other major ports in Britain.[95] Despite, or perhaps because of, its rapid growth, the trade in frozen meat between Great Britain and its antipodean colonies was not without hindrances. "Hurried and consequently careless stowing" of frozen cargo in preparation for its journey constituted a "chief danger."[96] The nature of the voyage between the antipodes and the North Atlantic itself posed a hazard. Ships from Australia and New Zealand spent between one and three months at sea, depending on means of motive power, much of which was "under an equatorial sun."[97] Equipment could (and did) fail, and obstacles were encountered during the journey. For instance, not only was the *SS Dunedin*, whose motive power was supplied the old-fashioned way—by wind—becalmed in the tropics, the ship's ventilation system became blocked by frost, threatening to compromise the cargo: "the cold air was not sufficiently 'tumbled about' amongst the carcasses." Only when the captain risked life and limb to fix it was the cargo secured.[98]

The risk of spoliation was significant at all points along this novel cold chain. The cargo of the *SS Protos* reportedly necessitated speedy cooking "because of the tendency to rapid decomposition."[99] But even before reaching London, frozen cargo was vulnerable to the vicissitudes of colonial climates. The perceived insalubrity of Australia's climate was thus in danger of being magnified by the technology of the trade. More problematic than the occasional failure of equipment, such as New Zealand's early shipments experienced, was the fact that any point of transfer for frozen carcasses—say, from railway car to ship, or from ship to storehouse—was an opportunity for thawing to occur, and thereby to do injury to the meat.[100] The Australian industry was especially prone to this vulnerability. Unlike New Zealand, where nearly all pastures were "far more favorably situated,"[101] located within easy distance of the colony's many ports, the bulk of Australia's flocks were grazed hundreds of dry, scorching miles from its ports. Many producers had to "drive their sheep perhaps 100 or 200 miles, and some of them even 300 miles, on foot," Bruce explained, "and then send them 200 miles by rail" to coastal freezing works. As a newspaper account of the Dunedin Chamber of Commerce meeting of 26 February 1881 proclaimed, "no reasonable man could suppose that meat slaughtered beyond the Blue Mountains"—the range dividing coastal New South Wales from its outback interior—"and sent to Sydney would, on arrival, be of a good color."[102] Such arduous journeys "deteriorat[ed] and wast[ed] the mutton,"[103] and consequently prevented "the meat taking first rank" or from "having any chance as a competitor with meat killed near the pasture."[104]

The alternative was to freeze inland and ship to Melbourne, Sydney, or other primary ports, but this option was nearly as problematic, as it left the frosty load vulnerable to total destruction should any mechanical failure or other impediment stall the cargo and leave it exposed to the punishing heat and sun of the continent. "It is during this transition that the success of the whole undertaking is most endangered," one observer noted, "for if the meat becomes at all thawed or softened in transit, the carcasses thus affected, when unshipped in the London Docks, present a most unpalatable appearance, being misshapen and discoloured, and are ... condemned ... as being unfit for food."[105] There was no quick fix for this dilemma, and it wasn't until the early years of the twentieth century that refrigeration and transport technology advanced sufficiently to allow for reasonably risk-free inland freezing in Australia.

But some of the lack of appeal of Australian meat appeared to be intrinsic. While Alexander Bruce, the chief inspector of stock for New South Wales, believed that the "ill-treatment and starvation" to which Australian sheep were subjected was responsible for the dark color of "our mutton,"[106] others suggested that the darkness of its meat was inherent. Skilling, an expert advisor to New Zealand's first freezing corporations, saw it as "a climatic effect" particular to "the meat of Victoria and New South Wales." In an estimation typical for the time, Skilling claimed that darkness was "inherent" in Australian meat because of the heat and aridity of the climate. As a more temperate, more Europe-like place, though, New Zealand was relatively immune from such danger: "Not only the meat of this colony, but the men and women too," Skilling promised, "were fresher and healthier-looking than those of the hotter climates of the sister colonies."[107] While the salubrity of New Zealand for man and sheep was celebrated, whether or not Australia was a "white man's country" preoccupied settler discourse, and in this case extended to its ovine co-colonists.[108]

Fraud and Prejudice

Some of the weak points in this cold commodity chain could be, and soon were, overcome. Warehouses for cold storage, for example—"among the most wondrous of recent developments in the river-side enterprises of London," according to a writer for *Chambers's Journal*[109]—were constructed in what another popularizer in Britain called a "topsy-turvy" manner.[110] Loading hatches were located near the roof, and carefully controlled chambers decreased in temperature as one descended toward the ground floor in order to prevent the "irruption of warm outer air into the cold storage chambers" as carcasses were loaded and unloaded.[111] Frozen meat, ovine or bovine, was often delivered "in a sail cloth," by means of "a crane with a very long jib" that could reach ships "lying at a considerable distance from the wharf," or by specially constructed "beef [or mutton] hoists."[112]

But not all obstacles to the trade could be so smoothly overcome, and perhaps the most serious impediment frozen colonial mutton faced was consumer prejudice. Britons marveled at the workings of refrigerating engines, and at the interior climate of cold stores. A journalist granted a tour of the *SS Orient*'s cold store in 1881 was taken with "the Arctic condition of the temperature," and the "white, snowy particles" that "had settled on the timbers and gathered on the wall till the whole had become touched with the heaviest of hoar frosts, and was sparkling at numberless points in the light of our lamp."[113] Another described the hold of the *SS Garrone* as "picturesque in the extreme."[114] Despite their fascination with the technology of frozen transport and storage, Britons also regarded early frozen cargoes with trepidation, and people had to be convinced that mutton that had "cropped pasture land 13,000 miles away, and been dead from six to nine months, or even longer" was good to eat.[115] Consumers worried about the effect of the freezing process on the "nutritive value" and tastiness of meat.[116] They particularly feared that the blood, and consequently nutritional value and flavor, would seep out of the meat during the thawing process, leaving it in a "dry and tasteless condition."[117]

Specific concern over the effect of the freezing process on the wholesomeness of colonial mutton was part of a wider unease with how, and increasingly, from where, the British got their nourishment in the nineteenth century. By the latter decades of the century, provisioning the domestic population of Great Britain had come to mean relying on "animal food" and grains produced elsewhere. Writing in the *Journal of the Royal Agricultural Society of England* in 1887, P. G. Craigie, secretary of the Central Chamber of Agriculture, remarked on this shift. Food was "still as imperative as ever for our fellow-subjects to find," he noted—no amount of progress could eliminate this basic fact of existence. But Britons no longer sought grains and chops produced exclusively on domestic acres, for as Craigie wrote, "world-wide is now the field whence it comes to our markets."[118] Coming to terms with these altered circumstances was difficult for a nation that prided itself on the consumption of fine meat, and on raising the animals that produced it. Britons initially resisted the colonial harvest of mutton and lamb, insisting on their preference for the home-grown article. "Englishmen prefer," wrote a contributor to the *Saturday Review* in 1881, "from taste or habit, English meat."[119]

An obvious way around "the extreme prejudice with which frozen meat was at first regarded"[120] was to undercut homegrown competition. As a writer for the *New Review* put it in 1897, "We do not eat Frozen Mutton and Refrigerated Beef because an Arctic temperature improves their flavour, or because the breeds and pasturage in other countries make better Meat than we can grow. We import them because they are cheap."[121] By the virtue of its availability, and because it could retail for several pence less per pound than home-grown mutton of comparable quality, colonial mutton found purchasers, even if not from among the most discerning epicures, in the early days of its trade.[122] Throughout the 1880s, colonial mutton sold for roughly one pence less per pound than home-grown, and by 1896 prime New Zealand mutton was two and a half pence less per pound than the top end of Britain's produce, while Australian mutton (alongside Argentinean) bottomed out at four and a half pence less.[123] Before American beef and Australasian mutton were widely available, meat was dear enough to limit its consumption by the working class, even though it made up a larger proportion of the laborer's diet in Britain than it did in most of Europe.[124] Consumers thus found it hard to resist such value, and the prejudice against frozen meat, as Leonard Lillingston observed, writing for *Good Words* magazine, was likely "mainly a middle-class one after all."[125]

Better-heeled purchasers were not as easily seduced by the great value of colonial mutton, but it found its way to the tables of the middle classes nonetheless. Despite assertions that discerning palates could tell the difference between locally raised and colonial imports, there was nothing to stop retailers selling colonial meat as home-grown. The best colonial meat, it was asserted again and again, though excellent, did not measure up to the very best home-grown, so that butchers selling the "bountiful supplies from the Antipodes" as Scotch and English mutton could make an extra profit.[126] As Lillingston noted rather astutely in 1879—the year of Australia's first shipments—"the British public would in theory have nothing to do with Australian mutton; but somebody appears to have eaten it, for the next year 17,275 carcases [sic] came into this country." This he attributed to the strong likelihood that "a great deal of it was sold as home fed, so that the consumer, through his own ignorance and folly, not only ate Colonial mutton against his wishes, but had to pay more than its market value."[127]

The extent of misrepresentation in marketing frozen meat from Australia and New Zealand, however, was probably more limited than anxious publications on the topic from the time suggest, not least because meat that has once been frozen presents a different appearance than meat which has never been frozen, offering an immediate visual cue to most purchasers as to the provenance (at least in broad geographical terms) of their supper.[128] Nevertheless, concern about actual and potential fraud was sufficient to convene a select committee in the House of Lords in 1893, on the marking of foreign meat, signaling the consequence of these matters to the body politic.[129] Various representatives of the trade paraded before the committee, giving evidence (sometimes contradictory) as to the quality and distinguishability of colonial versus home-grown meat, and to the persistence of fraud in London and provincial cities.[130] The committee's findings, which included the determination that consumers were "entitled to have English meat when they pay the price of English meat" regardless of any discrepancy in quality between foreign and domestic supply, upheld the connection between national identity and meat consumption.[131]

Even though the fraudulent sale of colonial meat was relatively insignificant in economic terms,[132] and even if the kind of worry that inspired the House of Lords Select Committee was overinflated, it remained culturally significant. Concerns about the misrepresentation of meat to British consumers spoke to precisely that centrality of meat to their daily lives and national identity. From the perspective of the colony, the misrepresentation of colonial meat constituted a "fraud upon the English consumer and New Zealand producer" alike.[133] The tendency to sell "New Zealand frozen mutton as prime English" was, in a way, a compliment to the quality of the colonial article.[134] Nevertheless, it left suppliers in New Zealand feeling they had been "'had,' 'robbed,' or 'swindled'" by British purveyors of their produce, a sentiment that the *Australasian Pastoralists' Review*—the premier agricultural journal for the region, and one that tended to be more sympathetic to the Australian contingent of its readership—remarked snidely, was "tenaciously cherished by many shippers of frozen meat in New Zealand."[135]

From the colonial vantage point, worse than swapping New Zealand mutton for prime English was the possibility that London butchers might sell "inferior English . . . as New Zealand,"[136] as this was damaging to New Zealand's reputation as well as to its profits. However, this charge was likely overblown, as the *Australasian Pastoralists' Review* was only too happy to point out. In 1892 the average price for carcasses of British sheep that had "met a fair death at the hands of the butchers" had yet to fall below—or even to the level of—those that "[came] out of refrigerating chambers." Until it did, this particular accusation was economically disadvantageous, and therefore unlikely to be made.[137] Though "fraudulent dealings"[138] with colonial mutton were problematic for both Australia and New Zealand, their frozen offerings were distinct enough to mean that the process played out differently according to specific colonial origin. Australian frozen meat was rated lower on the British market than New Zealand's, and thus it was more likely to be associated with the dreaded River Plate variety from Argentina, while the excellence of New Zealand's offerings made it vulnerable to being passed off as home-grown.

Display of frozen sheep carcasses outside the British New Zealand Meat Company, Christchurch, c. 1900–1920. Webb, Steffano, 1880–1967: Collection of negatives. Ref: 1/1-009113-G. Alexander Turnbull Library, Wellington, New Zealand.

The Australian colonies and New Zealand were "active and jealous rivals" in the frozen meat trade,[139] as in other arenas, but at least when it came to frozen mutton, New Zealand clearly led the trade. As Australians consequently had to hear "a good deal about New Zealand," the *Australasian Pastoralists' Review* complained, "they may be pardoned if they become a trifle weary of listening to the oft-told tale of the manner in which the sister colony emerged from her financial difficulties."[140] A more realistic and more damaging possibility was finding "La Plata mutton ticketed as New Zealand"[141] or as Australian, as was often the case. As "River Plate frozen mutton" was "far behind [that of] New Zealand in the matter of quality,"[142] the association with such an inferior article was a threat "to our good name."[143] It was likewise a threat to colonial profits for both Australian and New Zealand suppliers, and that, in the words of one Australian journalist, was where "the shoe pinches us."[144]

Native Colonials 127

Large flock of merino sheep, South Australia, c. 1900. National Library of Australia, nla.obj-144081153.

Part of what determined the differences in how fraud was perpetrated with regard to colonial meat on the British market came down to breed and climate, and to the various suitability of different types of sheep for colonial environments. The same challenges of climate that made Australia "not so suitable for killing and freezing" sheep put it at a disadvantage relative to its eastern neighbor.[145] Yet Australian mutton developed a reputation for being second-rate not only because its product lacked the "finishing" that came as a consequence of proximal pastures and abattoirs but also because merino mutton was as controversial in the 1880s as it had been in the 1810s. Cheap meat could always find a market in Britain, but no matter how good the price, Britons would almost always prefer the mutton of a "native" British breed over that of the merino.

With the advent of the trade in frozen mutton in 1880, Australian pastoralists quickly learned that they could not simply export mutton grown as a by-product of their wool industry to hungry, waiting consumers in Britain, and New Zealanders, observing the fortunes of their neighbors' mutton on the metropolitan markets, were quick to learn from such mistakes. The first frozen shipments from Australia were exclusively of merino mutton. And while Australians were certain that "it would be hard to beat for flavour a leg of mountain-fed Merino wether in good condition," Britons were not "sufficiently colonial" to agree.[146] A writer for the *Australasian Pastoralists' Review* believed that there was "no reason why the English taste should not be educated to a proper appreciation of merino mutton,"[147] but as the quintessential connoisseurs of quality in meat, they were not likely to be reeducated in matters of taste by their colonial cousins.

New Zealanders were happy to let their neighbors across the Tasman try: as one sensible producer there observed in 1892, "Merino wethers are unexcelled, but if the Home customer does not like them, why damage the trade. Oysters are good[,] so are eels, but I should be at starvation point before I would touch an eel, and I know of others who would have to leave the room if a dish of oysters were put on the table. Let the Australians and South Americans send this class. When they have educated the Home taste, then we can chip in."[148] Matters of taste were just that, and "it [did] not matter a rush what we in New Zealand may individually or even collectively consider the best mutton to eat." The customer was always right, and if pastoralists in both places wanted "to do the best we can out of the frozen meat trade, we must breed and ship mutton they like best in London, and pity their bad taste if we don't happen to agree with it."[149]

Cross Purposes

Prejudice against merino mutton presented an opportunity for New Zealand to capitalize on its more diverse colonial flocks. Even in the greener, more temperate parts of Australia, like Victoria, "when it [came] to competition as to mutton," wrote the *New Zealand Farmer*, "there is no comparison between the natural advantages of New Zealand."[150] New Zealanders were well aware that "though we can't come near Australia in the fineness and lightness in grease of our wool clip, we have all the advantage in climate for taking the lead in meat production."[151] Because of its wider range of regional environments and cooler, wetter, more temperate, and more varied climate, New Zealand was poised to take fuller advantage of the new opportunities of the frozen meat trade. The heftier carcasses of established longwool breeds in New Zealand, though fatty, were already better suited to the mutton trade than were the scrawny frames of merinos. While the longwool breeds predominated in the lowlands of New Zealand, other British breeds such as Southdowns and Shropshires could be found in smaller numbers, grazing the downs and foothills of Canterbury province in particular. These were champion mutton-makers, producing a pleasantly plump leg of mutton, but only a lightweight fleece of medium quality, making them (at least in the early days) relatively unpopular in the colonies.[152]

But shifting from an ovine population geared toward production for the wool market to one aimed at producing mutton came with its own challenges. As any good breeder knew, a sheep bred for wool did not necessarily produce good mutton. In fact, the case was more often quite the reverse. The relationship between the weight and texture of a fleece on the one hand and the carcass on the other was a problem that had occupied breeders in Great Britain since at least the late eighteenth century. Selection for one nearly always seemed to come at the expense of the other: if the merino's carcass was sway-backed and puny, Bakewell's improved Leicester was "deficient in wool."[153] In reformulating their flocks to suit the tables of metropolitan diners, producers were determined not to sacrifice wool for carcass. "Without a doubt," wrote Taylor White, the author of a two-part series on crossbreeding in the *New Zealand Farmer*, "wool as well as mutton must be kept going."[154] They therefore held wool and mutton, local environment and distant consumer demand, in the balance.

While their struggle to blend good meat and fine wool may have resembled the efforts of earlier generations of British breeders, the connection between locality and type was of a different character in colonial New Zealand than it was in Great Britain. As earlier chapters have shown, in the latter, types were understood to have evolved slowly over time, directed not only by the guiding hand of man but by the determining influence of climate, soil, and environment. To meet the imperatives of modern production in Great Britain, this tight connection between place and type had to be severed, or at the very least, weakened. But New Zealand was a *tabula rasa* for domesticated livestock, where types had to be *created* to suit localities as they were encountered.

In this, the intimacy between type and place that marked British breeding was a model for colonial breeders in New Zealand. "In England the natural habitat of the different breeds has been long since determined," declared an essayist for the *New Zealand Farmer*, "and we should in vain look for a Southdown in the fens of Lincolnshire, or a Lincoln on the chalk downs of Sussex or Hampshire."[155] In their own efforts, close attention to this issue was critical. "Everything depends on the kind of pasture a sheep is sustained on," wrote T. H. Anson, an early authority in sheep-breeding in the Canterbury region of New Zealand, in 1877. "Whether it will attain to a point as near perfection in carcass and wool bearing capabilities as possible; or, on the other hand, whether it degenerates every year in both" came down to the resonance between the land and the breed.[156]

As settlers learned the lay of their new land, they were also increasingly aware that the hills, dales, plains, and river valleys of New Zealand were like, but not *quite* like, "Home." For example, the seasons in New Zealand were just as topsy-turvy a version of Great Britain as a cold store was of a warehouse designed for regular articles. The *New Zealand Farmer* continually reminded its readership of this lest they, in their enthusiasm for the similarities between the "England of the South Pacific" and that of the North Atlantic, forgot their surroundings. September in New Zealand was "more like that of April in England as far as the weather goes,"[157] and November, too, was "different from the same month in the old country." This month might bring "to the minds of old people who have lived in London in the days of fog thick and yellow as pea-soup, candles lit and gas lamps at noon, and link boys with torches by daylight." In New Zealand, by contrast, it was "one of the best months in the year."[158]

For Samuel Butler, the British writer and malcontent who spent five years sheep-farming in New Zealand, Canterbury Province "reminded [him] much of Cambridgeshire" (provided a hazy atmosphere conveniently "obscure[d] the snowy range" of the Southern Alps, visible in the distance on clear days). Native cabbage trees, too, "which have a very tropical appearance," were "distinctive" enough to "characterise [Canterbury] as not English."[159] Just as the landscape was like, but not quite the same as, that of the "old country," as types of sheep bred for particular local conditions in the British Isles, British breeds were close, but not *quite* right for the colony. These uncannily familiar yet strange lands could be modified to a degree—they could be (and were) sown with English grasses, drained, or irrigated—but fundamental aspects of place such as altitude, exposure, climate, wetness, and temperature could be little modified.

Sheep, on the other hand, were far more malleable. Their character could be remolded to fit the land with more ease than the land could be refigured to suit the breed. "We must adapt our sheep to the character of land we possess," Anson exhorted other pastoralists, and the readiest, most effective way to do this was by crossbreeding disparate types in order to combine their traits in one variety.[160] Much as an earlier generation of breeders in Great Britain had, with great enthusiasm, crossbred local varieties in the name of improving existing breeds at the turn of the nineteenth century, colonial breeders in New Zealand employed the same methods, only they used them in an effort to produce the right combinations of characteristics for particular places. In New Zealand in the latter decades of the nineteenth century, that is, "improvement" indicated a desire to reconfigure existing breeds into new breeds "native" to the colony.

Crossbreeding had initially been undertaken in New Zealand as a way to maximize wool production. Early efforts to place the right type of sheep on the right type of pasture operated upon the theory of a cooperative "*chain of breeding*."[161] By crossbreeding merinos and longwooled breeds according to this principle, the properties of the merino—notably its fineness of wool—could cascade down from the high-country sheep stations, through the foothills and river valleys, becoming proportionally more dilute among the flocks in the approach to the lowlands and marshes. In the other direction, size, carcass weight, and weight of fleece—all markers of the long-wooled breeds—could climb gradually in diminishing proportion toward the highlands, the exclusive domain of the pure merino. In theory, this model meant that each sheep farmer could attain, by carefully calibrating his crossbreeding program, the right type of sheep for his pastures.

Even under the best execution of this principle, however, whatever type of sheep thus attained would have to be constantly re-created, as crossbred animals only breed true under very special circumstances. The more likely outcome was imperfect implementation of the "chain of breeding." A lack of "quality and lustre" could indicate, as in the case of the "Auckland district . . . that the settlers here have not got the right kind of cross," as the editor of the *New Zealand Farmer* opined in 1892.[162] "New Zealand" was as long a word when it came to regional variation as it was meteorologically,[163] and even in broad strokes, what worked for the South Island might not suit the North. As Taylor White noted, colonial sheepmen required "perhaps not an exactly similar sheep for both islands, but one to suit each district."[164]

In the "development of a breed," moreover, "Rusticus," writing for the *Australasian Pastoralists' Review*, advised breeders to make "the most of characteristics that are produced by the circumstances of their surroundings," rather than to "fight against Nature in trying to turn out animals similar to those grown in other districts or countries under quite different conditions."[165] Wherever pastoralists had the wrong kind of cross—or even simply ill-chosen parent types for producing crosses—constant infusions of "fresh blood" were the telltale sign that a "type [was] not suited to its surroundings, treatment, [and] pasture,"[166] and commentators had reason to lament the "very strong inclination on the part of many farmers to disregard the character of the land, and to be guided in their selection more by their fancy for a particular breed than by its suitability for the conditions under which it would have to be maintained."[167]

Exporting sheep for meat added yet another layer of complexity to the existing process, however imperfect. Put to work in service of the frozen meat trade, this theory of stratified production provided the means for breeders to recast their flocks as a compromise between local environments and metropolitan consumer demand. While colonial breeders felt sure that "the sheep farmers out here are naturally the best judges" of which "particular line of breeding" suited local conditions, they acknowledged that "London salesmen would, of course, know best what breed of sheep produced the mutton that sold for the highest price in their markets."[168] Opinion varied as to whether that breed was a Southdown, Shropshire, or something else, but all—in the metropole and colonies alike—agreed that the crossbred flocks of New Zealand "suit[ed] the taste of English purchasers"[169] and were "more highly esteemed in the English market than the merinos which Australia chiefly furnishes."[170]

A Bakewell for the Colonies

Crossbred sheep might have made a nice renewable product for the wool trade, and a good terminal one for the meat trade, but for breeders, the problem with crossbred sheep was that by their nature they produced instability and uncontrolled variation down the generations—the very thing that the establishment of pure breeds at the turn of the nineteenth century, as discussed in chapter 3, had worked so hard to forestall.[171] While a first-generation cross between a longwooled breed and a merino might dependably give rise to an animal combining the weighty fleece of the one with the fineness of the other, the offspring of *that* generation, depending on whether it was bred to a longwool or a merino, "naturally throws to the extremes" of one or another of its "parent stock," and the result could not be guaranteed as an improvement over the breed in question in its pure state.[172] This was especially so, John Roberts cautioned at the Intercolonial Stock Conference held in Wellington on 25 October 1892, wherever the "component parts of the blend" were "two such violent extremes as the merino and long-wool" as they were in New Zealand.[173] This was exactly the kind of cross between radically unlike types against which an earlier generation in Britain had warned, and it meant that "much difficulty, and more than ordinary difficulty," would attend any attempt to established a fixed cross between them. The very extremity of the cross, Roberts warned, "must of necessity tend towards frequent throwing back to the original strain, on one side or the other," and breeders ought "not to anticipate that the permanent establishment of the half-bred sheep in the colony as a distinct type will be ... easily secured."[174]

Nonetheless, in catering to the new productive imperatives of the frozen meat trade, and in response to the character of colonial pastures, sheep farmers in New Zealand needed a new breed, distinctly colonial but still capable of satisfying the tastes of the most discerning British consumers. "We require," Taylor White exhorted his compatriots, "to raise a new type suitable for New Zealand and the requirements of the meat-freezing industry."[175] "English bred sheep are not exactly what we want," breeders recognized as early as 1877, as they were apt to alter in some way in unfamiliar colonial environments, whether that meant succumbing to disease, failing to fatten, or growing coarse or rangy.[176] Rather, New Zealand wanted "some native breeds, which shall not need to go through a course of acclimatisation, nor be periodically reinforced by new blood imported for the purpose."[177] The challenge was how to achieve this, given the intrinsic instability of crossbred varieties. "We want a Bakewell to fix up a new type of sheep of permanent characteristics" was the call that sprang from the pages of the *New Zealand Farmer* in 1892. The "new type" should be "neither too large or the reverse, of a muscular or fleshy character, and one to arrive at the standard of weight and condition in eighteen months time."[178]

Tastes were changing in Britain, and it was important that this new breed should suit consumer preferences. "As mutton sheep," the large-framed longwools were "a thing of the past," declared "a Southland correspondent" in the *New Zealand Farmer*.[179] Though New Zealand fielded "smaller-bodied, shorter-legged, but better woolled ... Colonial types of Lincolns" than the "big-framed, upstanding English-bred Lincolns," even these were too large for the British market.[180] By the 1880s and 1890s such "mountain[s] of fat and tallow" had fallen from favor at "the tables in the Old Country."[181] Consumer preference in Great Britain had shifted toward leaner, more compact breeds like Hampshires and Shropshires, and other "fashionable" types lately "improved" by crossing with the Southdown.[182] Even the working classes had developed into "somewhat fastidious" consumers.[183] As a contributor to the *New Zealand Farmer* reported in 1892, "a greater mistake could not be made" than to assume "that the poorer class of people in England will eat the fat carcases of the Lincoln."[184] This individual had "personal experience amongst the agricultural, manufacturing, and mining population, and can say, that they positively refuse to buy fat mutton, if any choice is given them, because to them 'fat' simply spells 'waste.' "[185] Consequently, the "chief object" among New Zealand breeders, according to White, must be to "raise a medium sheep suitable for freezing" without fat "laid on in thick patches on the outside of the loins," as longwooled types were liable to accumulate.[186] Breeders aimed "to *increase the lean meat* in like ratio to the fat, in fact, to breed an *active, muscular* animal rather than a sluggard, or one almost dead from *fatty degeneration* of the system."[187] New Zealand's "native" breed was to be as fit and healthy as its can-do colonists.

Fixing the Cross

These characteristics were easy enough to produce in the first cross, but to replicate them over generations was no easy matter. To "fix" such characteristics, a Bakewell was just what the situation called for. While crossbreeding might produce a good terminal product for the freezer, breeders in New Zealand wanted a fixed and reliable type that could produce *generations* of "freezers," "the best stamp of ... breed for the freezing trade": they wanted, in essence, to replicate the advantageous points of a crossbred in a pure breed able to reproduce itself with consistency.[188]

Some parties doubted whether or not such a thing could even be achieved. The uncertainty and variability inherent in this process led to much anxiety. Breeders feared that efforts to establish a fixed cross would put the colony's flocks into a hopeless muddle, and that a lack of particularity would create an indeterminate horde of "mongrel-bred" sheep with no distinction, hence no predictability, in breeding.[189] A fellow by the name of Oliver, who was "a well-known sheep-farmer," for example, asserted in 1891 "that crossbreds have not sufficient fixity of type, and consequently continued breeding upon them will produce only worse and worse mongrels."[190] And when the Canterbury Agricultural and Pastoral Association introduced a category for crossbred sheep the same year, "no section in the sheep department of the Show occasioned more interest." The *New Zealand Farmer* reported that "exhibitors from north and south vied to show unbelievers that establishment of a 'crossbred breed' was not only possible, but had been actually accomplished."[191]

Efforts to establish a "permanent" or "fixed" cross, following the tried-and-true methods of their British predecessors, had begun as early as the 1860s. James Little, whose flocks in the 1890s were "the evolution of several crosses,"[192] began working toward an "inbred crossbreed" in Otago as early as 1868.[193] William Soltau Davidson, dissatisfied with the "uneven" nature of the "three-quarter-breds," determined to fix the half-bred type ("It was the half-bred sheep we wanted and nothing more or less") by intensive inbreeding of carefully selected crossbreds.[194] The flock he oversaw for the New Zealand and Australian Land Company, "kept perfectly pure and inbred"[195] since its inception in the 1870s, was eventually recognized as the oldest continuously bred flock of what came to be known as the Corriedale.[196] Others who left a less indelible mark on the new type of "native" colonial breed included Thomas Thatcher of Wanganui, whose "plucky venture"[197] had begun to garner attention in 1892. He had for some time "endeavour[ed] to establish a new type of sheep, combining the best qualities of the Merino and Lincoln," and his hybrid breed, "specially fitted for the frozen meat trade," was applauded in the pages of the *New Zealand Farmer* as a "demonstrat[ion] that the fusion of new blood has great advantages, the sheep being healthier, better woolled, and of excellent carcase proportions."[198]

The way to fuse "new blood" into a new breed was simply to apply the principles of inbreeding to carefully selected crossbred stock. Advocates of this method pointed out that "the greater number of the present *pure breeds* of British sheep have originated from the *crossing* of two or more of the original breeds in certain districts, for formerly each prescribed area within certain limits held its own distinct breed of sheep."[199] Even established breeds like "the Lincoln and Romney or Kent are both allowed to have been *improved* by Leicester blood."[200] Bakewell's own practice remained shrouded in the mists of uncertainty, but he, too, New Zealanders reassured themselves, likely infused the target of his improving zeal with genetics (or "blood") from another breed, subsequently inbreeding intensively to eliminate other than the desired characteristics.[201]

The trick for colonists in New Zealand was to select for "a carcase approach[ing] a square in every way,"[202] without sacrificing the lucrative high-quality wool for which their sheep were known. As Roberts had predicted, establishing this in a "permanent cross-bred flock" was a difficult feat.[203] As with any attempt to play Bakewell, the first step was to use superior foundation stock. Even if a farmer could not "go in for extra well-bred expensive animals," the ability "to pick the very best of the flock" for breeding was essential.[204] "Much more depends on the antecedents of family history than length of pedigree, or the appearance of the individual animal in question," the *Australasian Pastoralists' Review* cautioned. Sheep from well-bred, well-maintained flocks "will almost without exception, reproduce and forcibly transmit their qualities to the flocks they are mated with."[205]

Corriedale-cross sheep at Cheviot Hills Station, Canterbury, c. 1893.
Sinclair: Photographs of Cheviot County, chiefly of Cheviot Station.
Ref: 1/2-043179-F. Alexander Turnbull Library, Wellington, New Zealand.

Having such excellent material in hand, the next step was to "fuse" their "blood." Taylor White's recipe for establishing "a permanent cross-bred flock, or as we may call it a new variety," was to "work with three distinct varieties."[206] The efficacy of such a method he took "as an established fact."[207] After the first cross between a merino ewe and a longwooled ram, such as a Lincoln, the progeny would be bred to a third type—a Southdown, Leicester, Shropshire, or whatever kind the breeder desired "the offspring to most resemble."[208] This method had several virtues. By employing a greater range of types, it satisfied the partisans of several breeds, for whom the issue of which breed to use in the production of "freezers" was a matter of intense debate.[209] More importantly, White claimed it worked to minimize the tendency of the first cross to "throw back," and thereby provided something of a shortcut to "fixing" the cross.[210]

The real difficulty in a simple two-breed cross was that offspring tended to "vary greatly each from each when bred *inter se*," thus requiring "a matter of lengthy time and care before they can be bred to a uniform standard."[211] This was because "the first cross weakens the heredity or power of transmitting likeness." Having thus "thrown out" heredity, as White put it, "the third pure breed will more readily impress its likeness on the result of the first cross."[212] The "introduction of a third pure breed" seemed to smooth out the phenotype of a collection of animals by introducing another set of traits to its blended genotype. This added component "greatly impresses its type on the mongrel blood," and by "again using a pure sire of the last variety," a breeder might finally fix "a permanent and new type ... which will breed true within itself."[213]

The Southern Cross

Establishing a fixed cross was a lengthy process, whether it entailed fusing just two, or blending more breeds. The benefit of that expenditure of time was that it adapted the "new variety" to local conditions. Ultimately, a successfully fixed cross could be "ranked as a pure breed, true to type, and appropriate to the district and climate where raised."[214] This was the endgame for colonial breeders in New Zealand: the "inbred half-bred,"[215] in which was embodied "weight, substance, evenness of fleece, and symmetry combined in such a manner as is somewhat foreign to English sheep-breeders, who, it is known, consider mutton of paramount importance."[216] Though the Corriedale, as it came to be called several decades later, "[had] not the long ancestry of the principal sheep breeds of the present day," Holford admitted in 1928 "that it is an established breed, and that it breeds reasonably true to type has been adequately proved by hundreds of sheep-men in New Zealand and overseas."[217]

Indeed, if the aim was to produce a "native" New Zealand breed capable of satisfying distant consumer demand for "British" meat, the "inbred half-bred," as well as the crossbred mutton of various parentage that continued to fill the refrigerated holds of ships, was indeed "a triumph of the sheepbreeder's art."[218] So successfully had this project been that by the 1920s, "many people in [England] regularly [bought] 'Canterbury lamb' in the belief that the meat they are getting comes from the Canterbury district of Kent," rather than the Canterbury region of New Zealand.[219] The new Corriedale breed embodied these contradictions, serving as both New Zealand's own quintessential "native" breed and the ideal universal mutton-maker. New Zealand breeders were proud of how widely the breed had been exported—to Australia, South America, North America, Russia, and even Japan—at the same time that they celebrated its unique identity as a "native" New Zealand breed. No other breed was "so suited as the Corriedale [was] to ... the sheep lands of the Dominion," enthused the breed's first historian, G. H. Holford, or so capable of producing "magnificent mutton."[220] While acknowledging that "the British breeds of sheep are unsurpassed for the particular purpose for which they have been created," Holford gave credit to the Corriedale as an ingenious creation capable of taking advantage of the natural attributes of the New Worlds' grasslands. British breeds had taken root in New Zealand because of its affinities to "the Homeland," but equally strong affinities among "the upland sheep lands of the Dominion" made the Corriedale, as a "sheep bred to suit this class of country," eminently suited to commensurate pastures "in foreign lands," making it uniquely suited to new conditions of production, in which the distance between production and consumption expanded across the imperial stage.[221] As "New Zealand's own," the Corriedale breed served to validate the frozen meat trade in cultural as well as economic terms, a validation that extended even as far as the colony itself. As newcomers to a remote corner of the globe, for colonists in New Zealand, the creation and celebration of the Corriedale as a "native" breed authenticated their own presence there.

CHAPTER FIVE
A Universal Type

If, in the first half of the nineteenth century, the Hereford breed remained a strongly regional type, "native" to its corner of western England, by the 1880s, the breed had "approached the goal which has been so much coveted by [its] admirers," according to one journalist: "the first place among the bovine race," and it had done so by virtue of its "marvelous adaptability."[1] "Across the seas," enthused another writer for the *Livestock Journal*, "Hereford cattle have been, and still are, in great request. They seem better adapted to some parts of America . . . than either Shorthorns or Devons; and similar testimony has come from certain districts of Canada. There are numerous Hereford herds in the Australian colonies, and there are circumstances which cause this breed often to be preferred at the Antipodes to any other."[2] These included the ecological characteristics of much of Canada, the United States, and Australasia. In their hardiness, ability to forage, and phenotypic stability, Herefords were well suited to the extensive conditions of New World beef production in the latter half of the century.

But the circumstances favorable to the breed also included the technological proximity of metropole to colony afforded by refrigerated transport after 1875. The old truism "You are what you eat" is of special import, perhaps, in Great Britain, where at least since the eighteenth century, national identity has been pegged to consumption of meat, and where, as we have seen, worry over the nation's meat supply persisted throughout the nineteenth century.[3] Much as it had for the ovine biology of Britain's Australasian colonies, the shortfall between domestic supply and demand—combined with technologies of transport and preservation—refigured the bovine character of America. As the prairies increasingly grazed cattle destined for consumption across the pond, American stockmen worked to transform their herds of "native" cattle into beef fit for British consumption. They relied on British bloodstock—especially of the Shorthorn, Devon, Hereford, and Angus breeds—to "improve" the mongrel descendants of earlier waves of bovine imports, not least of the Spanish cattle that accompanied earlier waves of Iberian immigration to what became the American southwest. The demand for purebred Herefords in New York, Illinois, Michigan, and elsewhere in the United States—as well as

Ontario and British Columbia, New Zealand, New South Wales—was initially a boon to English breeders, bringing financial gain as well as the recognition they sought within the world of pedigree breeding in Britain: the Hereford's domestic reputation was, in many ways, forged abroad. But the long-term consequences of the late-nineteenth-century Hereford exodus were more equivocal. Materially, the demand for purebred bulls with which to "grade up" colonial and quasi-colonial stock allowed American and colonial breeders to establish their own genetic reserves, threatening British breeders' position of primacy within the global hierarchy of Hereford breeders, and ultimately challenging the character of the breed itself.

Type and Transposition

Toward the end of the nineteenth century, the sense that there was perhaps "no breed of cattle which has been exposed to so much opposition as the Hereford" persisted among its proponents in Great Britain.[4] An earlier generation had been more sanguine about the breed's prospects in the British Isles. After all, the oxen "for which the county of Hereford is famous"[5] were held in high repute among the graziers serving the London market, as we saw in chapter 3, and in the early years of the nineteenth century, it had appeared as though the Hereford breed was "spreading very fast, and will, in a few years, exhibit their white faces in almost every pasture in this Island."[6] Observers, like the one who signed himself T. S. in an 1808 letter to the *Agricultural Magazine*, were sure that "when this breed becomes more known," Herefords would have "the preference shewn them [that] they so justly merit."[7]

As the century progressed, though, no such preference was forthcoming. On the contrary, the Shorthorn breed continued to dominate Great Britain, geographically and rhetorically. By the latter decades of the nineteenth century this breed's popularity was, if anything, even higher than before. One particularly enthusiastic and high-minded proponent declared in 1885 the "Durham of Old England" to be England's "greatest combination of beef and milk," possessed of "a stately majesty of position" reminiscent "of 'Landseer's famous dog picture.'"[8] From a more practical perspective, the Shorthorn was "as prominent in numbers and power of good things as ever" in the 1880s,[9] and had spread so far and wide throughout the British Isles that it seemed poised to "monopolise the whole face of the country," according to a journalist for the *Livestock Journal and Fancier's Gazette*.[10]

A Shorthorn cow. From David Low, *Breeds of the Domestic Animals of the British Islands*, Volume 1, Plate 9. Rare Book & Manuscript Library, Columbia University in the City of New York.

By contrast, Herefords had come to suffer by association the want of prestige that characterized its breeders. True to their early nineteenth century origins, they remained the practical breed of tenant farmers, and even their more elevated proponents tended to be far less influential than the Shorthorn's. Champions of the breed chafed at the unfair state of things this produced. The "'Whitefaces' have had to contend against many difficulties," as one journalist put it, "not having been in the hands of monied men, but chiefly owned by tenant-farmers," so that "in competing for the front place amongst the bovine race, they have been without that support and influence which have been given to its most formidable opponent."[11] Such influence and support had created a demand for Shorthorns all out of proportion to the merit of the breed, according to one observer who believed that "fifty years hence, our 'craze' for them will be put on a par with the tulip craze [and the] South Sea Bubble craze."[12] In such a setting, where prestige drove the primary measure of value, what practical evidence Hereford enthusiasts mustered carried little weight. According to their proponents, Hereford cattle received more accolades than Shorthorns in the show ring, continually brought "better prices" in the auction ring, and had "always been a more economical feeder and grazier" than their rivals, yet they continued to rate second. Given their self-evident merit as a breed, "is it not strange," asked one agricultural journalist, "that the Press and the agricultural societies have not been more ready to encourage them?"[13]

Herefords were not as wildly popular as Shorthorns, but recognition beyond their traditional confines was slowly growing. If they had not managed to "extend themselves over the entire face of the land like the Shorthorn," it was clearly not a matter of the breed's inability to adapt to the different climates, environments, and systems of management of the British Isles.[14] On the contrary, evidence of the breed's adaptability abounded. In preparation for his widely read 1863 essay, breed historian and *Herd Book* editor Thomas Duckham collected testimony affirming the Hereford's transposability from breeders throughout the British Isles and farther afield.[15] Although Herefords had not expanded beyond their "native county"[16] in the same numbers as the "improved" Durham at that time, where they did seek greener pastures, the red-and-white cattle of western England proved to be highly amenable to a range of localities. The estate of R. W. Reynall in Westmeath, Ireland, for instance, was home to one of the oldest herds of Hereford cattle outside of England—bred *in situ* "for fully a hundred years."[17] (Reynall's "taste for high-class stock" was evidently as old as his herd, it having been "born with him," according to the *Livestock Journal*.)[18] There, Herefords "readily [became] acclimatized," they "retain[ed] their general character in every respect,"[19] and they even, Reynall claimed, "improve[d] from the moment they arrive in Ireland."[20] Other Irish breeders were equally enthusiastic. Herefords were "hardier than the Shorthorn, and more easily fattened" than other breeds, and in the view of Samuel Gilliland of Londonderry, "the best class of stock" he could "keep for the butcher."[21]

Testimony from elsewhere in the British Isles was hardly less effusive. "The praise ... bestowed upon the breed in the neighbourhood of Preston," in Lancashire—the old stronghold of Bakewellian improvement, and considered in the 1860s to be Shorthorn territory—was gratifying to one proponent.[22] And as Hereford breeders "push[ed] the Whitefaces further north," cattlemen in Scotland affirmed the type's value.[23] An Aberdeenshire breeder reported that "the Herefords are hardy and well adapted for this northern climate," thriving in situations where "the best shorthorns" proved too "delicate, and frequently died."[24] Closer to the breed's original stamping grounds, Herefords were in even higher repute. Duckham noted that they had "almost the exclusive possession not only of the county from whence they take their name, but also the nearby counties of Monmouth, Brecon, Radnor, and Salop," and were equally popular in Wales.[25] Their "hardiness on the mountain farms" was an asset in Cardiganshire, according to one estate agent, where "in this cold wet climate," Shorthorns did not "retain their character in a similar degree to the Herefords."[26] And on the coastal downs of Dorset, "so far from their being degenerated," Herefords were, as in Ireland, "much improved."[27]

The way the Hereford was said to "retain its character" across diverse regions and varied conditions was a key factor in the breed's modestly growing popularity in Britain. It indicated a fixity of traits that could only result from the purity born of hereditary isolation—whether produced by the accident of circumstance, or by the artifice of pedigree: the range of characteristics brought out under the new circumstances was narrow enough for the transposed breed to remain more or less as it had been bred, in terms of phenotype and behavior. In more practical terms, it meant that a breeder was likely to get what he bargained for—in the Hereford's case, an economical feeder and a hardy beef-producer—regardless of systems of management or idiosyncrasies of location. This attribute, however, existed in tension with the adaptive capabilities of a breed, without which a breed tended to languish or "degenerate" in a new setting. The inverse of fixity of character, the ability to adjust to new conditions and to thrive in foreign circumstances relied on the degree of variability in a breed's collectively embodied hereditary potential.

Thus, to be transposable, a breed had to strike just the right balance between adaptability and hereditary stability. Too much fixity and a breed would fail to flourish outside its habitual circumstances, but too much variability and it would evolve away from its characteristic type. The Shorthorn was known for the former, its inability to adapt to circumstances other than those for which it was bred (very intensive production) reflecting its long history as a closed breed. By contrast, when breeders in the nineteenth century praised the Hereford for the way in which it "retained its character," it was because it seemed remarkably ready to acclimatize in new conditions—soil, topography, dampness, dryness, luxuriance of feed or the reverse—without submitting entirely to local conditions. Even in the mountains of Wales, Herefords preserved their reputation as "kindly feeders"[28]; in Lancashire, their qualities of "quick feeding and the hardiness of their constitution."[29] In each new place, remote or near, their "admirable properties" endured, the breed neither "degenerating" as was so often the Shorthorn's lamentable tendency, nor evolving (so it then seemed) away from its desired traits.[30]

In Search of Greener Pastures

Evidence from within the United Kingdom of the Hereford's enduring qualities was encouraging, but the breed's remarkable transposability—that perfect balance between stability of type and adaptability to location—found its fullest expression beyond the shores of the British Isles. As conditions of production and consumption developed over the course of the nineteenth century, new opportunities for purebred British breeds arose in such faraway places as the North American West, Australia, New Zealand, and the independent states of South America, especially Argentina and Uruguay. These diverse environments and the more extensive system of husbandry there pursued tested the Hereford, offering a wider range of terrain for which the breed could prove its suitability.

These external opportunities resulted from the changing context of imperial meat production in the mid-nineteenth century. The same forces that impelled the production of Great Britain's mutton in New Zealand, the subject of chapter 4—industrialization, population growth, the rise of the middle class, steam transport—also induced Britons to seek out alternative sites of beef production. Population growth and the perception that domestic agricultural production had stagnated led to an atmosphere of anxiety surrounding the availability of "animal food" in mid-nineteenth-century Britain. Duckham, then editor of the *Herd Book of Hereford Cattle* and one of the breed's most energetic promoters, observed that "the dietary habits of a rapidly increasing and prosperous population daily extend the demand for meat on the one hand." But on the other hand, "the meat producing area" in Great Britain was "annually reduced by the construction of railways, opening of mines, establishment of manufactories, and the extension of cities and towns."[31] Agricultural production was being squeezed by the very forces that were driving up domestic demand, and by the 1860s, it seemed as though the situation had reached a tipping point.

Public panic over the possibility of a "meat famine" in mid-Victorian Britain encompassed perceived shortage of beef as well as sheep meat. The prospect of undersupply was particularly alarming "in a flesh-eating country like England"[32] where an adequate quantity of meat was "a very vital question," according to one expert agricultural journalist. It was true "that the English people desire beef along with their cabbage," but it was more than just a question of want: it was one of need, as well.[33] Meat-eating was believed to be essential "to the health and strength of the British Islanders," and "without the roast or the boil," proclaimed a writer for the *Livestock Journal*, "'John Bull' would soon become ... as sodden-headed as a diseased potato."[34]

The readiest solution was to import live cattle from foreign countries to make up for this deficit, lest the nation, "like poor old canine-kindly Mother Hubbard, [find] the cupboard bare."[35] Proximate sources were tapped first. By the 1860s, Ireland had become the "backbone of the English fatstock industry," and "Britain was said to be draining the Continent of every head of stock that could be spared."[36] In material terms, this amounted to as many as 700,000 cattle entering Great Britain through London, Liverpool, Hull, and other major port cities over the course of the 1860s. Still only 3 to 5 percent of the total meat consumption for Britain during this decade, it was nonetheless a marked rise in the consumption of imported meat from absolutely nil less than twenty-five years previously,[37] and the proportion of foreign meat consumed by Britons only continued to rise, already reaching 41 percent of total consumption by 1875.[38]

The Americas constituted a third source of foreign supply. Great Britain eventually drew shipments of live cattle, and later of chilled dead meat, from the United States, Canada, and South America, especially Argentina, but the United States was first in the establishment of a transatlantic cattle trade, and it remained the most consequential through the nineteenth century.[39] European live cattle imports outweighed their American counterparts until the early 1880s, but to British observers the United States brimmed with potential.[40] The vast grasslands of the prairies, so broad and fertile, seemed destined to "grow the cattle for the shambles of the world," while the existing rail, slaughter, and refrigeration infrastructure connecting the midwestern plains with eastern ports promised to supply beef in plenty, whether on the hoof or on the hook.[41] The American West and Midwest were a draw for private capital and joint stock companies, but they promised more than a safe return on British investment.[42] They were hailed as an opportunity to extend Britain's pastureland, metaphorically and economically, if not necessarily politically. Much like colonial New Zealand, the "natural advantages" of the prairies (which seemed to have a "special adaptation ... to stock-raising") suggested they could be turned into offshore loci for the production of British meat for British consumers.[43] In the right hands, the great inland sea of grass that stretched from Nebraska to the Canadian West could "bring into the market a part of the country hitherto regarded as barren and unfruitful."[44]

This was no less true of the Argentinian pampas, or the Canadian prairies (which actually were under Great Britain's political aegis), and in all cases, their potential rested on the same kind of technological proximity that brought the pasturelands of New Zealand's South Island into the productive orbit of London.[45] As the *Livestock Journal* boasted, Kansas was a mere day's rail journey from Denver or Kansas City, New York was "only sixty-two hours" distant, and "the traveller may breakfast at the Langham Hotel, London, in less than fourteen days after leaving."[46] The journey was no greater, in terms of time or distance, for cattle shipped from this region, only they, of course, would constitute rather than consume meals in London hotels. Without steam transport, the vast tracts of land where "the meadow grasses of England are congenial to the soil, and time will make the pasture as rich as any old pasture in England,"[47] would remain untapped, the haunts of Indians and bison rather than the productive landscape of cattle husbandry.[48]

By Hook or by Hoof

American producers soon began to make good on the potential of the prairies for both domestic consumption and export to Great Britain.[49] In the 1870s, American beef began to fill the decks and holds of the steamships that plied the North Atlantic, first on the hoof, then, in increasing volumes, on both hoof and hook.[50] The trade began in 1868 with an experimental shipment of eighty-eight live cattle that landed in Glasgow.[51] After a lapse of five years in which no live imports from North America reached Britain, it resumed again, growing exponentially in the early years, from 402 cattle landed at Liverpool and London in 1873 to over 200,000 ten years later.[52]

By 1885, these numbers translated to a "vast display of American beef" in London and Liverpool, an "astonishing feature in the Metropolitan markets."[53] Though "the quality varies more than in home bred sorts," observers applauded "how care in selection and classification helps in passing it into the shops of all classes of butchers, and into the households of all classes of people."[54] Against the enormous imports from America, "prime Scotch cattle ... [were] but drops in the bucket," according to the *Livestock Journal and Fancier's Gazette*.⁵⁵ Unlike the antipodean mutton trade, where the greater distance between colonial pastures and metropolitan tables precluded live shipment of sheep, live imports from America were preferable to beef imported chilled or frozen, which began to reach Great Britain in the autumn of 1875. Despite the logistical and hygienic advantages of the dead meat trade, live imports continued, largely because the purveyor of home-killed American beef did not "encounter the same consumer prejudice that he found against the chilled article."⁵⁶

But while the quantities and displays of American beef astonished observers in Great Britain, consumers found the quality of American imports less than satisfactory. Reports varied, but a journalist for the *Livestock Journal* spoke for many when he reported that much American beef—whether on the hoof or imported as dead meat—fell "far below our standard."⁵⁷ Once again, the strength of the connection between breed, quality, and the "discerning British palate" proved a challenge to the overseas expansion of Britain's pastures.⁵⁸ Such was the preference for British meat that, according to one industry expert, "a great many people ... would rather eat a tough steak from some old cow or bull, provided it was killed in this country, than a tender juicy and flavoursome meal of the primest [sic] pedigree-bred Argentine Ox."⁵⁹ While markets for meat of all kinds—fresh or frozen, foreign or home-grown—were diverse, and consumer tastes and purchasing power varied by class and income, the perception that Britons remained "exceptionally fastidious in [their] tastes for butcher meat" prevailed in the 1870s and 1880s.⁶⁰ Providing "the class of meat calculated to satisfy these tastes," moreover, took the same skill and forethought when it came to beef as it had in the case of mutton.⁶¹ A writer for *Chambers's Journal of Popular Literature, Science and Arts* spoke for many Britons when he wrote that "unless the Americans send first-rate meat, they need not send it at all."⁶²

American beef initially failed to meet the standard of British consumers partly because of the challenges presented by the hardships of a sea voyage. During an Atlantic crossing, the animals were almost guaranteed to lose weight and "condition." Even after regulations regarding adequate care and housing for the duration of the transatlantic voyage were put into place on both sides of the ocean, the risk of injury to live cargo or of fatality remained high.⁶³ The result was an almost unavoidable deterioration of the flesh of American beasts, before they even reached dockside shambles in Liverpool or London.⁶⁴

But the inferiority of American beef, by British calculation, was more than could be accounted for by a difficult oceanic crossing. Its substandard quality came down to breeding—or rather, to a lack of breeding. Britons were impressed and pleased by the sheer numbers of beasts the American cattle industry was able to send, but not with the obvious lack of refinement that marked these animals. Most of the United States' exported cattle came from west of the Mississippi River, and while eastern states in many cases had reasonably well-established pedigree breeding cultures, much of the rough-and-ready West did not.[65] In contrast to the stratified and highly developed world of pedigree cattle-breeding in Britain, a recent study notes that what motivated American cattle men "was not status but money."[66]

Combined with the expedience that the need to stock large tracts of land demanded,[67] this meant that the western *ranche* cattle were decidedly inferior to British breeds, and the "quality of their beef," by extension, was "naturally inferior."[68] Assorted types and "animals of uncertain or mixed ancestry"[69] predominated, many of which were what British observers called "unimproved natives"—often the feral or semiferal descendants of animals brought over with European settlers in prior centuries.[70] The worst among these were the "Texas type,"[71] which were nearly feral descendants of "beeves of the Spanish type" that had been roaming what is now the American Southwest and Mexico since the sixteenth century.[72] As beef cattle, they epitomized the inefficient, big-framed, unimproved "native" type. These animals were disfigured by their "long, spreading, half-turned-back horns," and hampered from making good beef by their "long legs, thin, lanky bod[ies], big, ill-put-together, ill-balanced bones ... thin thighs, and light waists."[73]

For James MacDonald, a Scottish expert in cattle breeds, American types were "too inferior" to British breeds to rate on the Home market.[74] Even compared with the more headstrong of the British, the mongrel herds of America were unrefined and ill mannered. To cattlemen accustomed to more placid breeds like the Shorthorn, they were positively wild. W. H. Sotham, visiting a farm in Abilene, Kansas, from Britain was happy to report that the fattening steers he saw there were "of fair marketable quality," but when he entered their paddock to get a better glimpse of them, "they all ran off in a body like deer."[75] And they were as uncouth in appearance as they were wild in behavior. Common American "prairie cattle" were "coarse, unimproved," and "ill-cared-for,"[76] while the backs of Texan cattle were "too truly of the Gothic style of structure to carry a large quantity of roasting

A Universal Type 149

beef."[77] According to one visitor to Wyoming, "no greater delusion could be indulged than to suppose that the Western ranche cattle are capable of producing the class of meat which brings a paying price" in Britain.[78] Not even British ingenuity was a match for the unseemly "native" cattle: "When fattened with the best of our skill," lamented the *Livestock Journal*, "their beef would still be of a secondary quality."[79]

Making the Grade

This prediction held for much of the nineteenth century: even as the volume of the trade grew, the prices American beef realized remained below prime Scottish, English, or Welsh beef. In the 1870s, American meat reportedly sold for one to two pence less per pound than "medium home sorts."[80] By the first years of the twentieth century, though, home-killed American and Canadian beef had surpassed second-quality English, and was close on the heels of prime English. In 1905, North American port-killed beef sold for 48 shillings per hundredweight, English first- and second-class beef claiming 50 shillings 6 pence and 46 shillings 6 pence, respectively—a difference of less than one pence per pound. By 1908, the gap had closed even further. The best North American beef now sold for 53 shillings 6 pence, English first class for 54 shillings, and second for 50 shillings 6 pence.[81]

The amelioration in quality that this actual and relative rise in price reflects followed vast and widespread efforts to "grade up" American cattle. Except for the existence of limited and well-contained collections of purebred herds in places like New York, Kentucky, and Ontario, throughout most of the continent, existing herds were little more than a hodge-podge of undistinguished types. By importing purebred British bulls to use on their "unimproved native" cows, North American stockowners could raise the quality of their beef, the sires imparting characteristics like size, bulk, and early maturity, and in so doing, bring the standard of American herds closer to that "refined, and what the Americans would call a very highly graded variety of cattle."[82] "Care and judicious breeding during the last three-quarters of a century" had brought about a manifest "improvement of cattle" in North America, according to one writer for the *Livestock Journal*.[83] As his colleague reported, "the people of Canada are becoming alive to the fact that it would be no difficult matter to double or treble the value and productiveness of their stock by improving the breeds and by bestowing due care upon them."[84] Much as their contemporaries in New Zealand recognized with respect to the frozen meat trade, North American stockmen saw the benefit of catering to the export market by taking seriously the connection that existed for British consumers between breed, quality, and discerning taste.

The combined efforts to raise the level of breeding in Canada, the United States, and South America together stimulated the growth of an enormous export market for pedigreed British bulls. Initial enthusiasm was for Shorthorn blood, which promised to increase the size and elevate the character of the mongrel herds of the Americas, not least because ranchers in the Mid- and far West were able to draw on existing purebred herds in Kentucky, Missouri, and Kansas.[85] Success on this front was forthcoming in America's Corn Belt, where the supplemental grains the Shorthorn needed to thrive were plentiful. In the well-settled East, the Shorthorn did much to improve American (and Canadian) cattle.[86] Numerically, it dominated recorded pedigrees in the United States, constituting 58 percent of all registered purebred cattle as late as 1884.[87] And by the mid-1880s, it seemed to have had a perceptible effect on the quality of American meat that reached British markets. A Shorthorn advocate writing for the *Farmer's Magazine* confirmed that the "value of the breed" overseas was "incalculable": "To judge from the American beef, alive and dead, which finds its way here, it will be only fair to suppose that the improvement in American cattle during that time must be almost entirely due to its agency. The leavening influence of this blood has spread over the greater part of an immense continent, and clothed its semi-wild cattle with marketable beef."[88] In so doing, it brought "every inferior and mongrel-bred kind of stock" found in America closer to that standard demanded by the "exacting" British palate.[89]

Despite their initial numeric dominance, Shorthorns proved to be too delicate for the extensive ranching system developing in the grasslands of western North America, Australia, and Argentina.[90] In the pastoral systems of the new worlds, large herds of cattle were often required to "out winter" in the mountains, or to survive the heat of a Texan or Australian summer with less attention than they would have received in Great Britain, and to roam relatively far for their feed.[91] And in the rougher conditions of new world beef farming, the breed's failings quickly came to light. Shorthorns required "rich food at all times, rich loamy soils, and to be well sheltered."[92] Abroad, too, they suffered from many of the same problems of overbreeding that they did in Britain. Just as British breeders had, in their great enthusiasm for "fashionable pedigrees," been induced to overlook form and constitution, the "Shorthorn mania" in the United States had generated similar hazards. But while British breeders so carefully balanced quantities of flesh upon delicate frames, Americans looked to "great size, regardless of symmetry, quality, and compactness," complained a writer for the *Farmer's Magazine*.[93] And they, too, fell under the sway of the pedigree. Naming one of the most famous and most controversial Shorthorn bloodlines, W. H. Sotham lamented that "a Duke, no matter how long and coarse his legs, how deep his flabby brisket, how thin his hide or slack his crops, or how extended his paunch, was in great request" among American stockmen.[94]

The problem with Shorthorns, moreover, was not particular to the Americas. In Queensland, Australia—one of the few dedicated beef-producing regions of the "southern continent"—one stockman had "great difficulty [in] getting bulls of a good hardy constitution, with the appearance of bulls about them, instead of a feminine look. With few exceptions," he complained, "'quality' seems to be the aim of breeders, which ends in 'delicacy,'" and like Shorthorns elsewhere, many of those in Australia were "bred from a long line of over-fed stock, and reared on over-stocked country," ultimately becoming "deficient in everything except quality."[95] The pinnacle of refinement in Great Britain, Shorthorns failed to thrive abroad, where the harsh conditions of extensive agro-pastoralism that characterized London's "global wests" made the "quality" that distinguished them at home a disadvantage.[96]

The Redcoats Are Coming

Unlike Shorthorns, Herefords appeared to display just the right proportion of fixity relative to adaptability for the realms of beef production opening up in the "neo-Europes" of nineteenth-century colonialism. Even if Shorthorns did "monopolise the whole face of the country in the British Isles to-morrow, and the Hereford breed [were] to be universally expelled," wrote the *Livestock Journal* in 1875, "there would still be ample room for the propagation of the latter in America and the British Colonies, many spacious tracts offering themselves both across the Atlantic and at the Antipodes, for which no other kind of stock are so well adapted."[97] Indeed, diversity of environment and extremity of climate seem to have posed little obstacle to the breed. In Jamaica, where "the temperature in the summer stands at 90 degrees in the shade," they withstood tropical heat with such ease that even "half-bred" animals were so superior, "you would scarcely suspect [them] as being any other than pure breds."[98] At the same time, the breed was well-suited to the long winters and "very changeable" climate of western Ontario: one Canadian enthusiast deemed them "most profitable for the western prairies."⁹⁹ They were also "taking firm root in South America," and "in the Australian Colonies" the Hereford breed was often "preferred at the Antipodes to any other."¹⁰⁰

Breeders in the United States found Herefords no less suited to their requirements than did those in Jamaica, Ontario, or Australia. In the 1880s, Herefords rapidly supplanted the Shorthorn breed as the preferred type for crossing or "grading up" the mongrel herds of native cattle in the United States.¹⁰¹ With Shorthorns, Herefords had been imported to the United States in the 1820s, but remained largely confined to the eastern states until the 1870s.¹⁰² By this time, the breed "had obtained a good footing" in Colorado, and according to a journalist writing in the early 1880s, they had "made more rapid progress in the public favor at the West in the last five years, than was ever made by any other breed of cattle in America in the same [amount of] time."¹⁰³ In "the western pasturelands of Nebraska, Wyoming, Western Kansas, Eastern Colorado, and western Texas," an "interest in Herefords" was "awakening," and by the 1880s, "our keen commercial cousins across the Atlantic," according to the (British) *Livestock Journal*, had "discovered that pure-bred Hereford bulls are the best sires for improving their native stock."¹⁰⁴ In short order, the "leavening influence" of Hereford "blood" was proving to be even more elevating than that of the Shorthorn.¹⁰⁵

Just why the Hereford—a breed "native to a temperate, well-watered English shire" eventually "[rose] to dominance" is, according to geographer Terry G. Jordan, "one of the unexplained mysteries" of cattle-ranching in the nineteenth century.¹⁰⁶ As Jordan notes, the ecologies of the grazing regions of the nineteenth century were diverse, ranging from "tropical savannas to subtropical pine barrens and midlatitude prairies, from fertile lowland plains to rugged mountain ranges, from rainy districts to semideserts."¹⁰⁷ In North America alone, cattle country in the latter decades of the nineteenth century comprised mountains and foothills, plains, grasslands, the semi-arid reaches of Texas, and the humid lowlands of the Carolinas. In Australia, cattle dominated the subtropical region of Queensland at the very north of the continent, and were also found among the colonies' flocks of sheep in the "semidesert" conditions that prevailed throughout.

In keeping with late-eighteenth-century opinion, and according to established notions about the ways in which environment shaped type, the suitability of the Hereford breed to such a range of conditions is indeed remarkable. But by the last quarter of the nineteenth century, livestock breeders perceived that the balance between the relative influences of heredity, environment, and human selection had shifted toward the latter.

Decades of "improvement"—careful management, "judicious" selective breeding—had crafted the Hereford, like other British breeds, into something more nearly approximating artifice than natural fact. Emphasis on the way Hereford cattle "retained their character" celebrated this apparent triumph of human will. Its success overseas in spite of, or perhaps because of, topographical diversity, climatic variance, and unfamiliar systems of husbandry was testament to its concurrent adaptability and genetic durability under nearly any circumstances.

The "badge" of the breed—its signature white face and red coat—bolstered its claim to universality in the 1870s and 1880s, just as it had supported breeders' assertions of purity of descent at mid-century.[108] The dominance of this trait gave visual confirmation that a Hereford bull had been "at work" in a herd—an especially valuable attribute prior to the widespread adoption of barbed wire in the late 1880s.[109] Selective breeding requires "running" certain fertile female animals with a given male. Only in conditions where fencing permitted such sectioning of a herd (or flock, as the case may be) could paternity, and therefore pedigree, be verified. Even a fence was not always a guarantee. As Harriet Ritvo notes, "stolen kisses" and broken fences had bedeviled British pedigrees since their inception.[110] Barbed wire alleviated some of this uncertainty, but even after the 1880s, much American pastureland remained unfenced, making containment of a portion of a herd for selective mating impractical. It also rendered close observation at this crucial phase of the productive cycle, and subsequently during calving, equally improbably, meaning that cattlemen would have little assurance that a pedigree bull purchased for the purposes of upgrading an undistinguished herd had in fact performed his duties—unless, that is, the bull in question was a white-faced bull who "color-marked" all of his offspring with the same trait, regardless of the coat color of the dam.[111] In the extensive conditions of beef production that sprang up in the new worlds, the white face thus became a most important proof of parentage for grade beef cattle, and a guarantee of the elevating influence of improved British blood.

Importantly, the Hereford breed excelled at ameliorating the quality of the beef New World cattle produced without sacrificing the hardiness or self-sufficiency of rangeland herds so necessary to their survival. In this regard, Australia offered perhaps the truest test for the breed. While the Shorthorn was no match for the hot and arid "brown continent," the Hereford's particular combination of hardiness and docility appealed to Australian breeders. As an "active, yet most domestic animal," Herefords were able, even in the scorching heat of Australia, to endure long marches with equanimity.[112] Australian cattlemen "invariably found the Herefords the best travellers," the difference between them and other breeds—especially the Shorthorn—"being most noticeable in hot weather." As Robert Archer, a Queensland breeder, recounted, "On a hot summer's day, even when taken with the greatest care, a mob of Shorthorn bulls will have their tongues out in the first half-mile; and on two separate occasions I have known a Shorthorn bull to lie down and die from the heat, although they had been carefully driven."[113] By contrast, Archer had "never known a Hereford bull to knock up from the heat."[114] At the same time, Australian producers appreciated the Hereford's docility. The breed was more placid and not given to the freaks of temper that afflicted Devons and Shorthorns "down under."[115] Though many "accuse[d] the Herefords of rowdiness," the "worst night smashes" Archer had heard of "[had] occurred in mobs of Shorthorns," while Devons "want[ed] well watching," and if anyone doubted it, a Tooloombah breeder named Beardmore invited them to "come to my yard when branding and take a hand at catching the Devon calves, and his shins will soon convince him."[116]

Yet there were apparent limits to the Hereford's suitability to all climes and all places. Morocco, for instance, confounded the breed. When Edwyn Arkwright, the brother of the prominent Herefordshire breeder John Hungerford Arkwright, tried to establish a herd of Hereford cattle in the coastal Mediterranean region of Saifia in the 1880s, his cattle dropped like flies. In July of 1885, he took up "a reluctant pen to announce the decease of the 3rd and last Hereford heifer Primrose."[117] This was eighteen months after the "confinement and decease of poor Cowslip,"[118] and Curly, who "calved a month early." Both of the calves in question died, and Curly herself followed them "a month later of anemia."[119] Finally, in 1889, Edwyn Arkwright was forced to confess to his brother "that I cannot get on with our Herefords!"[120]

How much of his failure was due to North African conditions, and how much to his own ineptitude, was (and remains) difficult to determine. The challenges of the Moroccan climate were no doubt severe, but Edwyn faced criticism from neighbors and acquaintances "that we are not feeding them properly," and the fact that he asked his more knowledgeable brother in 1884, "How many days are Herefords supposed to be in Calf [?]" does not inspire retrospective confidence in his abilities.[121] Nevertheless, whenever Herefords failed to thrive, or succumbed to extremes of climate, apparent failings could be (and usually were) laid at the feet of the breeder. Simply stated, according to a writer for the *Farmer's Magazine* in 1881, "if the Herefords do not win easily" against the Shorthorn breed in America, "it must be the breeders' fault."[122] That was part of the beauty of such fluid nineteenth-century ideas about heredity and environment. The relative pull of any of the three forces that together made a type—environment, heredity, and human influence—was in constant flux, and could therefore excuse any evidence that might put a favored breed in a bad light. As the Hereford's reputation for suitability across a range of places and conditions grew, human error in particular was an increasingly convenient straw man for champions of the breed, helping to maintain its reputation for transposability.

In spite of occasional setbacks, the dominance of Hereford cattle across the seas was growing. Herefords "quickly adjusted to range conditions and established their lasting popularity as range cattle," according to modern commentary.[123] In the expansive productive regime that grew up in the Americas and Australasia, this spoke as much to its productive ability as to the breed's climatic adaptability. New world cattle industries increasingly specialized in grass-fed, as opposed to grain- or stall-fed, beef production—a purpose for which the Hereford had always excelled beyond the Shorthorn. "No beef," wrote a contributor to the *Farmer's Magazine*, "is better eating than that of the Hereford when fully ripe off the grass."[124]

Paradoxically, because of these productive requirements, as ever more diverse and distant tracts of land opened up, the range of breeds deemed fit to stock them narrowed. If the presiding doctrine of the late eighteenth century had been that "every soil has its own stock," in the geographically expanded pastoral context of the 1870s and 1880s, it had become more a case of one (or sometimes two or three) breed(s) fit all.[125] Climate and environment remained salient, even if they were no longer the sole factor in determining type, and rhetorically, breeders still exhorted their fellows to "seek for the cattle that suit the country."[126] Breeders ought, in the opinion of one Queensland cattleman, to "notice the country and the feed, and then purchase accordingly."[127] Great Britain offered an instructive case in matching type to locality for colonial Australia: "If in a small country like England different breeds suit different counties," this fellow asked, "how is it possible that one breed can suit a continent like Australia?"[128] Great size, however, did not equate to great diversity. Despite its greater landmass, Australia's microclimates varied less than those of the geographically small, but meteorologically diverse, British Isles. Still, the breeds that pastoralists sought as the century progressed were more and more often limited to the Shorthorn, the Hereford, and sometimes the Devon breeds. And given the Hereford's record for universal adaptability, it was fast becoming the one-size-fits-all type for grasslands beef production.

Spare the Knife, Spoil the Herd

All this generated a brisk market in Great Britain for Hereford bulls bred for export. Not even the global agricultural depression that "made itself felt...throughout the civilised world" in the mid-1880s could check the demand for "thoroughbred Hereford cattle" in America, according to the *Livestock Journal*.[129] The extent of the trade at its height went unrecorded, but it was undoubtedly great. Prior to 1880, fewer than 200 Herefords were exported to the United States; between 1880 and 1886, the volume was as many as 4,000.[130] The Hereford Herd Book Society itself only began careful record-keeping after the trade peaked, but recorded a total of 1,259 cattle exported from Britain between 1890 and 1901.[131] A significant drop from the 1880-86 estimated high water mark, the volume of trade in the 1890s was still consequential. Almost half of these were destined for the United States (663), South America as a whole taking the next largest proportion (495), with sundry exportations to Europe (27), Canada (20), South Africa (14), Australia (12), New Zealand (7), and the West Indies (4).[132]

The influx of Hereford cattle to the Americas was astonishing. Purchasers acting on behalf of foreign cattlemen often bought large consignments of Herefords. A lot of 100 bulls, for example, was purchased "in England for shipment to the grazing regions of Buenos Ayres" in 1883.[133] North Americans, too, made "large purchases" in Britain, proving to one British journalist that their "Transatlantic neighbours" had taken up Herefords "in earnest."[134] Hereford bulls used to grade up the herds of the American West were also filtered through eastern states.[135] A single breeder, O. H. Nelson, who (according to one effusive historian) "did more than anyone else to establish the breed on the Great Plains," brought 10,000 Herefords to western Texas in the 1880s alone, only some of which came direct from Britain.[136] In 1883, Charles Goodnight, "the famous Panhandle rancher," took twenty-five bulls, 625 cows, and 400 calves—all "excellent-quality Herefords"—in a single purchase from an Illinois breeder.[137]

For breeders and observers on both sides of the Atlantic, enthusiasm for Herefords abroad contributed to their growing reputation in Great Britain. As one journalist noted, the closer the breed "approached the goal which has been so much coveted by their admirers"—that of world

"SIR BARTLE FRERE"(6682) at 11 MONTHS. OLD.
Bred by and the property of Mr T.J.Carwardine Stockton Bury. Leominster.

Portrait of Sir Bartle Frere, an influential bull exported to America. From the *Herd Book of Hereford Cattle*, Volume 12, 1881.

domination—the more breeders in the United Kingdom took note: "In America, Australia and other parts of the world the position of the breed is assured, and its ultimate complete triumph in Great Britain is only a question of time."[138] And as they fanned out over North and South America and Australasia, "indications that the merits of the Herefords are now being more recognised in [their] own country" abounded.[139] Although the breed's geographic reach and numerical strength within the British Isles never matched its rhetorical profile, the "estimation in which this famous stock [was] held" in places like Buenos Aires, to which 159 cattle were exported between 1890 and 1901, aided in the Hereford's "becoming fully appreciated in England."[140]

But the impact of the breed's overseas popularity in Great Britain was more than simply a rhetorical elevation in its status. The thriving export market of the 1870s and 1880s (what one historian recently called the "Yankee boom")[141] had a material effect, if not so much on the breed's distribution within the British Isles, then on how breeding was carried out and monitored there. Demonstrating and safeguarding the purity of the breed had been a challenge and a source of controversy throughout the century, as we saw in chapter 3, and it remained so in the context of sharp and persistent overseas demand for Hereford bulls. British breeders soon ran up against what appeared to be a fundamental limit to pedigree breeding: "you cannot make a new pure bred Hereford except by breeding with what we have now got."[142] A seemingly reasonable observation of the point (and practice) of pedigree breeding, the truth of this statement in fact depended very much on how purity was defined. Moreover, it proved to be one that Hereford men were willing to overlook in their rush to satisfy the demands of overseas buyers.

So sudden and so intense was the demand for pedigree Hereford bulls to use in upgrading New World herds—first in the United States, and then primarily in South America—that supply in Great Britain very quickly came to seem insufficient to meet the requirements of overseas buyers. British breeders did all they could to satisfy the heavy demand, but supply lagged. Writing for the *Farmer's Magazine*, W. H. Sotham reported on the "numerous car-loads" of blooded bulls that were "constantly going to Texas, Colorado, Wyoming, and Montana, to improve the stock on the plains."[143] Increasingly, the corrective effect of this influx of improved blood was "plainly visible in the stock now coming from either of these cattle producing regions."[144] More and more, the "Grade Herefords" reaching the British market met with approval, but the "demand for thoroughbred bulls" continued to exceed available supply in Britain and, Sotham feared, it would to fall short "until more are bred."[145]

Artificial selection had the most dampening effect on the availability of bulls for export. One of the fundamental principles of selective breeding was to allow only superior specimens of each sex to procreate. Most commonly, this was enforced by drafting off unsuitable females from the breeding herd, and by castrating substandard males. At the height of the "rage for Herefords,"[146] the market for "Hereford bulls of all kinds" was, according to the *Livestock Journal*, "more or less remunerative."[147] The "use of the knife" to castrate substandard male specimens "[was] consequently limited."[148] In their haste to meet the demands of North and South American buyers, British breeders allowed an "excessive rear of bulls," and failed to "alter" many bull calves as they ought to have.[149]

As the largest market by far for Hereford bulls, American breeders did not hesitate to criticize this trend. They noted, and remarked upon, the deleterious overall effect of such laxity in selection. Much to their disappointment, "inferior specimens" were continually put up for sale.[150] An American correspondent to the *Livestock Journal* wrote "somewhat strongly as to the character of some of last year's exportations." Buyers from the United States had noticed "the bad condition and defective pedigrees of several of the cattle." It would, the correspondent declared, "be much better to steer many of the bull-calves, and put them upon the market for the butcher."[151] And as another contributor to the same publication noted, the "rush has been so keen and fast for Herefords" that some breeders were "unable either to name their animals or give their pedigree."[152]

The Stain of Blood

Sparing use of the knife was one way to make "new purebred" Herefords, however inferior. Maintaining flexibility in the pedigree system was another method that breeders in Great Britain were apparently as willing to pursue. Because purity itself was a construct, so too were the standards that governed it. In the 1880s, these standards were exclusively the rules for entry into the *Herd Book of Hereford Cattle*. After 1878, when the Hereford Herd Book Society assumed control over the *Herd Book*, for any new animal, breeders were required to demonstrate at least four generations of "named Hereford blood" on the side of the sire, and three on that of the dam.[153] Prior to this time, while the *Herd Book* had remained in private hands—first those of Thomas Campbell Eyton (who edited the *Herd Book* between 1846 and 1853), and then Thomas Duckham (1854 to 1878)—little other than unwillingness to submit genealogical information for a given animal was a bar to entry. Truculence on the part of breeders acted as some form of selectivity: both editors struggled to eliminate the "confusion among pedigrees,"[154] but continued noncompliance hindered their efforts, and selectivity based only on a lack of breeder cooperation was a poor standard, indeed.

The stewardship of the Hereford Herd Book Society promised better governance, but even their regulations looked sturdier on paper than they were in practice. The official purposes of the new society in 1878 included not only the intent to collect and publish the life histories of all "thoroughbred" Hereford cattle, but to "verify ... information relating to the pedigrees of Hereford Cattle," and of equal importance, "to investigate cases of doubtful and suspected pedigrees."[155] Even on the record, the society admitted that "so careless" had many breeders been "in the matter of pedigrees," it was "impossible, without serious injury to the usefulness of the work, to adhere with stringency" to the rules of their own making.[156]

At just the time at which British breeders sought ways to increase the supply of pedigreed animals available for sale and exportation, the growing popularity of the breed overseas put pressure on them to tighten entry to their herd book. The greatest pressure to patrol the quality of British pedigrees came from American breeders. When the *American Hereford Record* was established in 1880, it had "made the English Herd Book a standard," which is to say that "such animals as were admitted" to the *Herd Book of Hereford Cattle* in Britain and subsequently exported to the United States, were also admitted without question to the *Record*.[157] Despite the confidence in British standards this implied, the *American Hereford Record*'s parent organization, the American Hereford Cattle Breeders' Association, seemed conflicted about the degree to which the genealogies of their own animals relied on those of British Herefords. While they celebrated the fact that "many of the pedigrees" contained in the *Record* could be "trace[d] through English herds for one hundred years,"[158] they warned that this very antiquity made "absolute correctness" in their own volume hardly possible, and chose to lay it out expressly "so that time will remedy these defects."[159]

But time, rather than remedy these defects, only intensified concern over the apparent laxity of British pedigrees. Dissatisfied with "Short pedigrees"[160]—those animals for whom their breeders were unable to demonstrate the requisite crosses, and whose entries in the *Herd Book of Hereford Cattle* had thus been marked with a dagger[161]—the prominent American Hereford breeder T. L. Miller pleaded for greater care and attention on the part of British breeders. Such was the "present and prospective importance of the Hereford interest," he argued in a letter to a British counterpart, that it was paramount "that our rules should be so framed as to give the guarantee of purity of breeding."[162] The most energetic and powerful voice of the American "Hereford interest," Miller felt that careful "examinations of pedigrees" was "a duty that the managers of the Herd Book owe to the breeders," and tempering his criticism of British standards, suggested that more attention to documentation would suffice to eliminate "the majority of short pedigrees."[163]

When the American Association tightened its own rules for entry to the *American Hereford Record* in 1883, the apparent discrepancies between the management of purity on either side of the Atlantic became even more problematic. In February of that year, the American Hereford Cattle Breeders' Association resolved to clamp down on deficient genealogies. New entries to the *Record* "and their produce" would be required to "show *first*—For the Sires of such animals, *5 Sires*; or for the dams of such animal, *4 Dams*."[164] Given the nature of the export trade, and the blood ties between Hereford cattle in Britain and America, the American association charged Miller with the task of "confer[ring] with your Society," as he wrote in an official missive to S. W. Urwick, secretary to the Hereford Herd Book Society, "with a view to securing uniform action as to the rules governing the Entry of Animals in your Herd Book, and the American Hereford record."[165] Under the association's new resolution, the first thirteen volumes of the English *Herd Book* would remain "as a standard, unless there should appear to be errors or fraud," but "all animals not entered in the first 13 Volumes of your Herd Book" would have to meet the new, elevated requirements, just like any American-born Hereford.[166]

Miller's proposal was proffered as a polite but firm invitation to join the American Hereford Cattle Breeders' Association in their effort to "frame" their rules "so as to give the guarantee of purity," but the association's terms left the British Herd Book Society with little choice.[167] "If however the English Hereford Herd Book Society elect to enter animals showing only two dams without giving any explanation," Miller remarked to John Hungerford Arkwright, "they of course have the right to do it. But it will subject them to criticism that it would be well to avoid," he cautioned, and moreover, registering "animals [that] may be 1/2 and 3/4 bloods" as purebred would "have a tendency to lessen the value of full blood animals."[168] The implied threat was, of course, the loss of the American export market: as far as the American Association was concerned, British breeders could either measure up to their new standards or do without their business.

Unsurprisingly, this came as something of an affront to breeders in Great Britain, and they resisted such pressure. According to Joseph Russell Bailey, a member of the Hereford Herd Book Society's editing committee for much of the 1880s and 1890s, Hereford men in Britain had enough trouble as it was in meeting the existing standards. "Tinkering with the rules" so shortly after having made them would be, in his estimation, "a distinct disadvantage" to the society's authority in the eyes of Hereford breeders, and a "disturb[ance] to public confidence."[169] Entry to the *Herd Book* had only recently ceased to be ad hoc, and "there are some breeders (Herbert Cranshaw for instance)," Bailey remarked to J. H. Arkwright in 1883, "who are missing . . . this pedigree until they reach the third or fourth cross."[170] If the society "made this rule more stringent," Bailey argued, "it would be disheartening for anyone in this position." Given that there had been "a lapse of only four years since the rule was made," he felt it was "almost a breach of faith on the part of the Society" to attempt to so tighten entry to the *Herd Book*.[171]

Even more to the point, four crosses of "named Hereford blood" seemed perfectly sufficient to Bailey and to his compatriots.[172] For example, after four crosses the descendants of a cross between a Shorthorn and a Hereford would be "1/16 Shorthorn blood, 15/16 Hereford which," Bailey argued, "should be pure enough for anything." Even the third cross produced an animal that was "7/8 pure," "beyond which," by his calculation, "the stain of blood is not carried." In a rare explicit reference to the racial typing of his own species current at the time, Bailey continued, if "in America a man with this amount of black blood would I believe be considered absolutely white," then so, too, ought a Hereford be considered absolutely pure.[173]

Globetrotting on Four Hooves

Of course, what counted as white or not in nineteenth-century America varied according to time, place, local norms, and changing legal regulations, but was as arbitrary as the measures that governed the purity of "thoroughbred" livestock. Nothing about this comparison seems to have struck Bailey—or Arkwright, for that matter—as absurd, inappropriate, or in any way irrelevant. That the American Hereford Cattle Breeders' Association wished to enact regulations for pedigreed cattle more strict than those imposed upon the people of their own land merely seemed unjust and insulting to British bloodstock and its breeders. The 1881 letter in which Bailey literally drew parallels between pure and crossbred cattle, and the racial classification that ruled contemporaneously in some parts of the United States is a rare but nonetheless compelling reminder of what is more often an unstated aspect of this subject: that discourse surrounding type could be as easily applied to people as to animals.[174]

It is also an indication of the significance of the "Volume 13 rule," which marks a watershed in the history of the breed. The late 1880s were both the dawn of the heyday of Hereford cattle and the moment at which the influence of its original breeders began to wane. The Hereford had succeeded in transcending its county origins and become a breed of nearly global proportions. But after the loss of the American market, the export trade in Hereford cattle dwindled. Demand in South America—tied, as in the case of New Zealand's frozen mutton trade, to the growth of the refrigerated shipping industry—rose, and exports to Canada continued more or less unabated, but even combined, these sources could not make up for the American demand at its peak.[175] British Hereford breeders would have another moment in the sun in the early postwar years, when Britain's Milk Marketing Board made the semen of Hereford bulls widely available and wildly popular for use on dairy herds through artificial insemination,[176] but after the institution of the infamous "Volume 13 rule," they never again benefited from the material and symbolic benefits that came from their earlier monopoly over the breed. Meanwhile, the Hereford stock exported from Britain generated and regenerated, proliferated and altered beyond the control of its original breeders in Great Britain. In their haste to profit from the overseas demand for Hereford cattle, British breeders had forgotten the lesson of Robert Bakewell, and allowed the "genetic template" embodied in their breeding stock to escape their grasp.[177] The consequences for the status of this "native" breed would be profound.

Conclusion
The Return of the Native Breed

In 1892, an article in the British *Agricultural Gazette* (reprinted in the *New Zealand Farmer*) boasted that Britain had recently "attained the leading place as a breeding and distributing centre" for sheep stock.[1] While other places—Spain, Australia, New Zealand—had the advantage of climate over Great Britain, according to this author, "careful attention to breeding and general management," supported by Britain's "favourable condition in regard to commerce," had allowed it to rise to the top of the world of livestock breeding. Such a position was hard won, but easily lost: Spain's precipitous fall from glory in the eighteenth century after centuries of dominance as the world's repository of merino sheep was a handy lesson if Britons were tempted to "get a little elevated over our position."[2] Moreover, British sheep breeders only had to recall that their "best blood" (indeed, their only blood) was not, at least by some measures, British, having come from elsewhere: all writers on the subject, according to the *Agricultural Gazette*, "point[ed] with one finger to Asia" as the locus of *Ovis aires*'s domestication.[3]

Furthermore, by the 1890s, some of Britain's "best blood" had already left the British Isles. Australasia had "drawn on us for breeding Leicester, Lincoln, Southdown, Shropshire, and Cheviot blood."[4] Canada, the United States, Patagonia, the Falkland Islands, and Argentina, too, had siphoned off their share. Many of these places had drawn upon Britain's reserves of thoroughbred cattle as well, as seen in chapter 5. These regions, according to this commentator, needed watching. New South Wales, Queensland, and New Zealand were "making rapid strides towards the lead in breeding and distributing" sheep, and had "already begun to send out blood towards the Cape, South America, and California."[5] Indeed, New Zealand cherished just such hopes of supplying its neighbors with stud stock. Only quarantine stood in the way, according to a report of the colony's Livestock Committee in 1891, and were this hindrance removed, "in a few years New Zealand would become the source from which the Australian colonies would draw their supplies of highly-bred sheep and cattle."[6]

Despite mild anxiety over the gestures of such colonial and quasi-colonial upstarts in this direction, Britain remained largely secure in its position as the locus of "the production of animals of the highest class" until at least the mid-twentieth century.[7] By then, however, changing global patterns of production motivated new developments in breeding away from long-standing British aims. High yields and high productivity became the watchwords of postwar farming, not just in Britain but all over the world, and many of the British types, developed for a different set of market imperatives in the nineteenth century, fell from favor. As "continental" varieties moved into British pastures, they also increasingly took the place of British types in their erstwhile overseas territory. Even where "British" breeds continued to dominate as, for example, the Hereford breed did in the North and South American cattle industries, they did so largely outside the grasp and influence of British breeders. If, in the nineteenth century, adjusting to the fact that "world-wide is now the field whence [foodstuffs] comes to our markets" had been a difficult pill to swallow, in the latter twentieth century, the fact that the "blood" that supplied the world's livestock industry was increasingly sourced from elsewhere was, for the former champion, more bitter medicine still.[8]

Far from being sought after by pastoralists around the world, many British breeds had tumbled so far from their former glory that some of them faced extinction even in their native land. Holsteins and Friesians increasingly took the place of Shorthorns and Ayrshires in the milking stall; and beef breeds like Limousins, Charolais, and Belgian Blues replaced Devons and Lincoln Reds in the paddock, while Dutch Texels and Finnish Landrace sheep supplanted Cotswolds, Ryelands, and Leicesters. The high yield of these specialized continental breeds, whether in milk or meat, constituted a threat to the more generalized, multipurpose nature of native British breeds.[9] The very glory days of British livestock husbandry—the nineteenth century—formerly so secure, were now themselves in danger of eclipse, and their signature breeds at as much risk as the rare, marginal, "primitive" types like the Soay.

Toward the close of the twentieth century, these groundswells in global livestock husbandry induced yet another shift in the meaning of "native" breeds as the Rare Breeds Survival Trust (RBST) announced the establishment of "a new category of endangered breeds."[10] Called "Native British" breeds, the RBST defined their new targets of conservation as the "pure original types" of British livestock.[11] Initially, the trust focused on "genuine traditional cattle."[12] The native bovines on which the organization first focused its attention were what Peter King, the organization's field officer responsible for education and outreach, described as the "direct descendants of those animals registered in the first herd books," quite an inclusive category given the relative novelty of published herd books.[13] Compared with commercial types, these animals were "slower maturing," hardier, "more efficient in converting low quality forage" to beef, and in general, "respond[ed] well to non-intensive production methods."[14] The threat these "foundation-type animals" faced was not necessarily extinction, which had seemed to loom over primitive breeds like Soay sheep a generation earlier, but a more insidious demise resulting instead from "introgression."[15]

A term borrowed from population genetics, introgression referred to the process by which a population's genotype could be infiltrated by an exogenous gene or genes, and thereby irrevocably altered. This described, in modern technical terms, the genetic effect of crossing, what nineteenth-century experts like David Low called "intermixture."[16] Anytime two types were interbred, whether intentionally or not, their hereditary material blended, their genotypes altered. The term itself carries no necessary moral charge. Whether an "official [or] 'unofficial'" act, it simply connotes change within the collective genetic potential of a type, and often, by extension, in the phenotypes of a group's constituent individuals.[17] This process had been the making of some of Britain's best-known breeds—witness the "improved" Shorthorn, the product of crossing introduced Dutch cattle with the old Yorkshire breed in the early eighteenth century, or almost any of Britain's longwooled breeds in the nineteenth century, nearly all of which benefited from Leicester "blood."[18]

But these breeds had been produced in the spirit of improvement, the opposite of the RBST's conservationist drive, and King argued that alterations could be taken too far: "change and 'improvement' are not an evil as long as a breed retains its fundamental characteristics," but unfortunately, this "[was] not always the case."[19] Too often, it seemed, stockbreeders succumbed to the imperatives (and temptations) of modern production. Seduced by "rapid short-term financial gain," they were "attracted by the drastic measure of introducing an entirely different breed" to that which was already in their possession.[20] The "ever increasing influence of continental breeds" like the tall and hefty French breed, the Charolais, meant that without conservationist vigilance, purebred British types, many of which had been carefully maintained by "handful[s] of breeders who have faithfully continued to breed the pure original types," might be bred out of existence, or simply abandoned.[21] The trust hoped that by singling out those "genuine traditional" types of cattle that "deserve recognition," they could forestall the loss of any more breeds.[22]

The trust's initiative to conserve "native" breeds was thus a response to—even an effort to prevent—the passing of the nation's agricultural glory days. And although the idea of a "native" breed was far from a new formulation on the part of the RBST, the trust's impulse to conserve them was an appropriation of the term under novel circumstances. Conserving British livestock heritage and resources had been the crux of the RBST's agenda since its inception, but the issue of native belonging had entered into this imperative primarily with reference to very marginal archaic, seemingly aboriginal breeds of sheep and cattle. Pursuing the conservation of the purebred types of the nineteenth century in connection with their nativeness to the British Isles was a new—and important—articulation of the issue. This shift acknowledged that the place of Great Britain in the world of livestock breeding had altered: it was no longer "the stud farm of the world," and its native breeds had slipped from the position of dominance they had enjoyed in days past.[23]

But the choice of poster breed complicated both the narrative of former glory and the conservationist response it engendered. The trust named the "original Hereford type" as the inspiration for their new classification—a native British breed whose genetic makeup seemed to be in such danger of alteration from introgression that "modern stock [bore] little resemblance to the original breed."[24] According to the cutting-edge DNA analysis used to verify the plight of the breed in 1996, the trust was able to identify "just 350 pure British Herefords not showing signs of alien influence" out of over 2,000 cattle tested.[25] These findings confirmed what a small group of breeders had been observing since the late 1970s and 1980s: that Herefords in Britain were changing dramatically—not for the better, and perhaps irreversibly—as a result of imported foreign genetics. The culprits, however, were not the continental giants used widely among British beef and dairy herds since their first introduction in 1961,[26] but rather bulls of the Hereford's own breed, bred and raised in Canada and the United States, and reimported to Britain as part of the same general enthusiasm for "taller, leaner, bigger, more uniform" cattle that had brought the Charolais and Limosin across the Channel.[27]

The postcolonial bovine immigrants that threatened "pure English" Herefords were larger than their British counterparts, and they reached maturity faster. More than this, they looked and behaved differently: their markings were perceptibly different from those of the English type, and their constitution was more delicate than that of Hereford cattle bred always and only in Britain. Where "English" Herefords were "broad in the beam, short in the leg, horned, and with the mellow skin and propensity to put

A Traditional Hereford cow, bred by Les Cook, Cambridgeshire, United Kingdom. Photo by the author, 2009.

on weight from the roughest fare as had their ancestors before them," some of the *arriviste* animals were "so tall and narrow as to scarcely qualify for the title of beef animals."[28] Nevertheless, many British breeders greeted them enthusiastically, and used these Canadian bulls widely, hoping to impart their size and precocity to their own herds. Appreciation for the imported cattle, however, was not universal, and a small a subset of breeders observed these developments with increasing trepidation. They were not yet convinced of the superiority of the newly arrived Herefords, and as they observed the effects of Canadian genetics on English herds, they began to discern a threat to English pedigrees. The characteristics of English Herefords seemed at odds with those of the Canadians, and the outcome of mixing the two were Herefords that did not look like they should, at least to proponents of the English type. This cautious minority began piecemeal efforts at preservation, and as individual action coalesced into growing recognition of the differences between Canadian and "pure English" bloodlines, the English type gained status as a distinct variety, first by the Hereford Cattle Society in 1995, and then, in 1996, by the RBST.[29]

While the high degree of introgression revealed by genetic testing suggested that the Traditional Hereford was in a grave state, by other measures it was an unusual choice as the poster breed for the RBST's new initiative. In the first place, both the breed's history as a "native" type and its relationship to the British Isles as a whole were complicated. The very nativeness of Hereford cattle, even to their own county, had been contested throughout the nineteenth century, as chapter 3 described. And its identity as British, moreover, was arguably weaker than other more recognizably national types. White Park cattle, for instance, were more obviously ancient and evidently aboriginal than the Hereford (which made them both more native and more British), as was the White Park's supposed domesticated off-shoot, the British White breed, which the RBST adopted as its logo.[30] Even the Hereford's old rival, the Shorthorn—which dominated the national herd for dairy and beef until the 1930s—was in a way a more logical choice, having effectively been born British.[31] That is to say, unlike the Hereford, which was a "county breed" par excellence, the Shorthorn's profile since the beginning of the nineteenth century was decidedly a national one, indicated by both its nearly universal distribution within the United Kingdom and its nomenclature, which included terms denoting locality—(the "improved" Durham breed, the "improved" Yorkshire breed)—but which was dominated by a categorical name based on a physical trait rather than its geographical origins.

Moreover, the Hereford's road to national recognition had taken place in no small part on colonial and American soil. As chapter 5 demonstrated, by the 1980s and 1990s, when "native" British breeds fell into the crosshairs of the RBST, a substantial proportion of the Hereford's last 100 years of development had taken place beyond the shores of the British Isles, in the context of British imperial, or quasi-imperial, expansion. Unlike a number of other British breeds—Longhorns, Lincoln Reds, or even Devon cattle, for instance—whose circulation beyond the United Kingdom was more limited, Herefords were exported so widely and in such great numbers that global circulation became a hallmark of the breed. This was a necessary condition for the breed's eventual conservation: without the widespread exportation of Hereford cattle in the nineteenth century, there would have been no "foreign" Hereford genetics to threaten the "pure English" type. But it also became a defining characteristic of the breed. Even proponents of the traditional variety acknowledge and celebrate the breed's global dom-

The entrance to the Hereford Cattle Society offices boasting the breed's global domination, 6 Offa Street, Hereford, United Kingdom. Photo by the author, 2009.

inance. Without a trace of irony, Traditional Herefords are touted as both the quintessential native British breed and the ultimate world traveler. At once "pure English," descended from "entirely *British Bloodlines*," and "the universal beef breed," Traditional Herefords, in an unapologetic nod to the erstwhile empire, are described as "the breed on which the sun never sets."[32]

Yet the course of conservation for the Traditional Hereford has worked to eliminate, or at least to implicitly obscure, the imperial legacy of the breed. In defending "pure English Herefords" from the tidal wave of returning postcolonial Hereford pedigrees that threatened to overwhelm indigenous English pedigrees, proponents of this type redrew the bounds of the Hereford breed, carving out a breed-within-a-breed that could be construed and defended as native. In doing so, they implicitly privileged environmental factors over shared genetic roots: time spent outside Britain and in the hands of unfamiliar breeders, not common origins in nineteenth-century British stock, came to define the reimported former colonial varieties. Identifying generations spent on foreign lands, in unfamiliar climates, and in the hands of unfamiliar breeders as the primary defining factors—rather than shared genetic and historical roots—thus redefined certain Hereford cattle returned to their erstwhile native land in such a way as to deny their claim, such as it might be, to regional and national belonging within Britain.

At the same time, Traditional Herefords were set apart as an environmentally and culturally autochthonous sub-breed, or breed-within-a-breed, celebrated as *the* ultimate English breed—perfectly adapted to its climate, its environment, its particular system of production. Emphasizing the connection between breed, beef, and patrimony, "it may not be too long," the RBST's Peter King wrote in 1996 of their native British cattle initiative, "before British pastures are [again] filled with beef cattle that not only are part of our history and heritage, but also produce the quality of meat that the discerning British palate will appreciate."[33] In drawing such connections, and by implicitly disinheriting modern pedigrees, efforts to conserve the Traditional Hereford also disinherited the legacy of Britain's imperial history—the reciprocal return of the erstwhile colonial, in this case clad in the red coat (and white face) of the Hereford breed.[34]

IN THE ARC of this story can be seen the changing significance and consequences of so-called native breeds. Central to the development of Britain's status at the top of the world of stud stock in the nineteenth century, regional and local types "native" to parts of Great Britain were first the raw material used by agricultural "improvers" to forge such titans of improved livestock as the New Leicester Longwool and the Shorthorn breed of cattle. Connection to place—to soil, climate, terrain, temperature, and conditions—was important: it was the first measure of distinction in a type, conferring the identity and "character" essential to its recognition as a breed. Later, it leant antiquity, purity of descent, and an elevated status relative to other breeds. As Great Britain's "favourable condition in regard to commerce" developed over the course of the nineteenth century, and in particular as "space and time [were] annihilated" under an expanding system of steam transport,[35] drawing together the four corners of the empire, "the tinge of origin" that adhered to types of stock was an important measure of new lands.[36] Their various fates added to basic observation, teaching colonial breeders about the commensurability of unfamiliar lands, the intimacy of understanding thus developing over time as breeders molded existing types to new lands, and the lands to their types.

Too stubborn a connection to locality marked a breed as a loser in the nineteenth century: this was the heyday of ovine and bovine transposition, where types and breeds were moved about from one distant place to another, in great numbers, and with much enthusiasm. A breed like the Hereford, able to "retain its character" while simultaneously prospering across a range of climates and conditions, succeeded where more strongly localized types were not even tried, and where a more finicky type like the Shorthorn languished. Transposability had to be carefully balanced by "character," and both were produced by means of selective methods over the course of the nineteenth century.

Of course, there were limits to transposability, and they were most starkly encountered in the colonies, where unfamiliar lands demanded adjustment on the part of breeds. New types like the Corriedale were forged according to tried-and-true means and aims: produced by cross- and in-breeding methods, this breed, as discussed in chapter 4, claimed to embody the elusive goal of an earlier generation of British improvers—an English sheep in Spanish wool. The aim and outcome of these efforts were a hybrid type, calibrated to both distant consumer demand and the realities of colonial topographies. New breeds like this, and existing types modified for colonial conditions, performed an essential function. Aided by the technologies of industrial transport, especially refrigeration, they satisfied the burgeoning appetites of Britain's "urban carnivores" for British meat produced by the "ghost acres" of the colonies.[37]

In this, the link between pursuits agricultural and pastoral, and the good of the nation that underpinned the rhetoric of improvement in the early nineteenth century, was given free range on an imperial scale. The connections between power and prosperity, population and sustenance, patriotic pastoralism and national security moved beyond Britain's shores. The trade in frozen colonial mutton that arose in the last decade of the nineteenth century rescued "the home food supply in the shape of meat" from dependence on "America and the Continent of Europe."[38] Of no small consequence for a people for whom the consumption of animal flesh was central to their collective identity, such commerce and traffic were understood to be mutually beneficial: "British ships could not only bring meat from the colonies," but take in return manufactured goods, thereby "conferring mutual advantage."[39]

Such notions of rosy symbiosis obscured a more complicated, more conflicted reality. Almost any imperial traffic rested on a legacy of the violence of conquest, and the frozen meat trade was no exception. A breed like the Corriedale, with its genetically British roots and its colonial "character," aspired to an embodiment of the mutual advantages of colonialism, but its significance as a both a colonial and a "native" breed reveals the darker side of imperial ties. While nativeness with respect to livestock could and did mean many things, in the colonies it was handmaiden to colonial dispossession. European colonialism—across time and place and nearly without exception—relied on the dispossession of indigenous peoples. The particulars varied, but rarely (if ever) was violent conquest not part of the process. In New Zealand, the process of Maori dispossession—what Evelyn Stokes calls a "tenurial revolution" in New Zealand—was piecemeal, periodically peaceful, but also violent, the bloodiest and most sustained moment of conflict the Land Wars of the 1860s.[40] To call a breed "native" in such a setting was thus a political claim as well as an environmental one: whether consciously or not, the establishment of "native" colonial breeds rhetorically bolstered claims to imperial dominion.

Imperialism, as many scholars remind us, though, was a two-way street, and as with so many other aspects of the empire, the colonial breeds eventually came home to roost.[41] Anxiety over the purity and the future of the "Traditional" Hereford breed coincided not only with the disintegration of the former empire but also with the successive waves of immigration of former colonial subjects to Great Britain.[42] Concern over the impact of postcolonial creole breeds upon native British breeds of cattle mirrored disquietude over the effects of widespread human immigration on British society and culture.[43] Unlike in an earlier moment, where nineteenth-century breeders in Britain felt free to openly discuss the parallels between breed and race, between the animal and the human conditions, such connections remained submerged in this case. As the more racially and culturally diverse population of Britain evolved in the late twentieth century, the realm of breed conservation remained an unusual discursive space in which conversations about English purity and nativeness continued to take place relatively unapologetically. Here, claims to native belonging or indigeneity were rolled out in an opposing fashion to their deployment in colonial settings. Where the plasticity of the concept—and of the animals themselves—supported colonial breeders working to establish their claim to foreign lands in the face of people with obvious prior claim, in postimperial Britain, that very possibility for evolution and adaptation (or creolization) was the ground upon which to deny the right of belonging and the claims of blood. Where Herefords in the nineteenth century "claimed each other as one family," breeders in the late twentieth century maintained that the legacy of expansion had pulled them asunder.[44] In a last stand for Britain's superiority as the stud stock capital of the world, defenders of the Traditional Hereford, and breed conservationists in general, redefined "native" as the repository of crucial national heritage.

The decision to thus cleave the Hereford family tree seemed to settle the question J. C. Hindson posed in 1974 to his fellow breed conservationists at the RBST, and which framed this work's introduction: "is a Soay still a Soay after 25 generations in the South of England?"[45] More than twenty years later, with their choice of the Traditional Hereford as the emblem of British "native" breeds, the RBST answered Hindson in the negative. Over time, their decision suggested, transposition amounted to a difference in kind. By this measure, a Soay was *not* a Soay after twenty-five generations in the South of England. A broader view, however, suggests that the matter is not so easily put to rest. Flexibility in the meaning of a breed itself, and especially in the signification of "native" as a label, ensures to the contrary that the answer to such a question will always be equivocal. What "native" meant at a particular time or in a particular place was highly contingent, and like the domestic species to which it was (and continues to be) applied, it was subject to great variation. The history of Britain's "native" breeds affirms this.

Notes

Introduction

1. Campbell, "St Kilda and Its Sheep," 28.
2. Clutton-Brock et al., "Sheep of St Kilda," 24, 25–29. See also Campbell, "St Kilda and Its Sheep," 30–31. Campbell notes that Soay sheep are not amenable to herding, and "must be run down and captured individually," although whether this is attributable to their prehistoric character or simply to their more recent prolonged semiferal existence must be a matter of debate. Ibid., 31. See also Harman, *Isle Called Hirte*, 190–93.
3. Earliest estimates have Soays brought to Britain circa 4500 B.C.E. by an early wave of human migrants. Fraser Darling, "Foreword," in *Island Survivors*, x, italics in original; Clutton-Brock, Pemberton, and Coulson, "Sheep of St Kilda," 28, 29. These sheep have had the run of Soay, their original islet, and Hirta, the largest of St Kilda's composite parts, since the early 1930s, when the archipelago's last permanent human inhabitants voluntarily evacuated. Long before this, though, successive changes to human occupation and the introduction of more modern breeds of sheep had progressively marginalized the breed, eventually isolating them on Soay, where they continued to subsist largely beyond the reach of human interference, thus inadvertently preserving a set of intriguing archaic attributes. The Norse were the next to arrive with their own ovine domesticates, a type known as the Hebridean. A millennium or so later, in the mid-nineteenth century, the St Kildans replaced these old Norse sheep (now known as Boreray sheep for the small island to which remnant herds were confined) with Cheviot or improved black-faced breeds popular in Scotland. Harman, *Isle Called Hirte*, 192.
4. The human population of St Kilda was always small, never exceeding 200 at the highest estimate. Prior to the emigration of a full one-third of its population to Australia in 1852, the number of St Kildans sat between 100 and 110. The population continued to decline, especially under the dampening demographic effect of World War I, until 1930 when the remaining inhabitants, unable to sustain their island economy, evacuated. Harman, *Isle Called Hirte*, 124–41, 134; Richards, *From Hirta to Port Phillip*, 110.
5. R. M. Lockley, "Wild Viking Sheep of Soay," *Country Life* 77 (10 March 1960), 509. See also Morton Boyd and Jewell, "Soay Sheep and Their Environment," 360.
6. Notably populations of sheep from the Isle of Man and the Orkney Islands.
7. Contemporary estimates are based on the number of breeding females in existence, which the Rare Breeds Survival Trust (RBST) estimated to be between 900 and 1,500 in 2016. RBST, "Watchlist 2016," http://www.rbst.org.uk/%252FOur-Work%252FResource-Library%252FWatchlists%252FWatchlist-2016; RBST, "RBST

Fact Sheet—Soay," http://www.rbst.org.uk/Rare-and-Native-Breeds/Sheep/Soay. Cf. RBST, "Guidelines for Acceptance onto the Rare Breeds Survival Trust Watchlist," http://www.rbst.org.uk/Our-Work/Watchlist/About-the-Watchlist. Earlier estimates ranged only as high as between 650 and 700 in 1948. Morton Boyd, "Introduction," in *Island Survivors*, 2.

8. J. C. Hindson, "Questions on Trust Policy," *The Ark*, no. 1, December 1974, 18.

9. Ibid.

10. These fall into a category of lands that Crosby has called "neo-Europes" for their ecological homology with Europe. Crosby, *Ecological Imperialism*, esp. 2–6 and 147–51.

11. Pearce, *Berkshire*, 46. See also Wood, "Sheep Breeders' View," 230.

12. Halifax, "Foreword," in *Britain Can Breed It*, 5.

13. Franklin, *Dolly Mixtures*, 53.

14. Abigail Woods explores some of the impetus to breed livestock for higher yields in "Breeding Cows, Maximising Milk."

15. Halifax, "Foreword," 4.

16. Virginia Anderson, *Creatures of Empire*, 3.

17. See Crosby, *Ecological Imperialism*; Crosby, *Columbian Exchange*; John McNeill, *Mosquito Empires*; Jones, *Rationalizing Epidemics*. For a more sweeping view of epidemiology in history, see William McNeill, *Plagues and Peoples*.

18. Virginia Anderson, *Creatures of Empire*; Benjamin Breen, "'Elks Are Our Horses'"; Cronon, *Changes in the Land*; Melville, *Plague of Sheep*.

19. Crosby, "Virgin Soil Epidemics"; Jones, "Virgin Soils Revisited."

20. Melville, *Plague of Sheep*. Melville's analysis of the Valle de Mezquital has subsequently come under revision from historical geographers who assert that the ecological condition of central Mexico must take into account the more complex history of pre-Columbian and colonial land use, not only the impact of introduced ungulates. See, for example, Sluyter, "Making of the Myth in Postcolonial Development," 377–401.

21. Flannery, *Future Eaters*, 41, 42.

22. Cronon, *Changes in the Land*.

23. Virginia Anderson, *Creatures of Empire*, esp. 7.

24. See Brooking and Pawson, *Seeds of Empire*; Muir, *Broken Promise of Agricultural Progress*; Dunlap, *Nature and the English Diaspora*; Beattie, O'Gorman, and Henry, *Climate, Science, and Colonization*. For a similar argument made in the context of the Caribbean, see McCook, "New-Columbian Exchange."

25. Bradford, *Bradford's History*, 166.

26. Valenze, *Milk*, 144; ibid., 139–45 for cattle in the Americas more generally.

27. Ritvo, *Platypus and Mermaid*.

28. Derry, *Ontario's Cattle Kingdom*; Derry, *Bred for Perfection*; Peden, "Pastoralism and the Transformation of the Open Grasslands"; Sayre, *Ranching, Endangered Species, and Urbanization*.

29. For the twentieth century, the opportunity to pursue breeds as a tool of analysis is even more pronounced. For a recent example, see for instance, Saraiva, *Fascist Pigs*, esp. chaps. 4 and 6.

30. For example, Isenberg, *Destruction of the Bison*.

31. Cronon, *Nature's Metropolis*, esp. 225–30.

32. Dunlap, *Nature and the English Diaspora*. For the transformation of human identities within the context of empire, see Catherine Hall, *Cultures of Empire*; Burton and Kennedy, *How Empire Shaped Us*.

33. Harrison, *Climates & Constitutions*; Chaplin, "Creoles in British America"; Chaplin, *Subject Matter*.

34. For instance, see Farland, "Modernist Versions of Pastoral." For an example of a primary source writing against this tendency, see Jefferson, *Notes on the State of Virginia*.

35. For example, T. H. Breen, "Empire of Goods"; Cook, *Matters of Exchange*; Magee and Thompson, *Empire and Globalisation*.

36. Mintz, *Sweetness and Power*.

37. Hall, *Civilising Subjects*. See also Hall and Rose, *At Home with the Empire*; Wilson, *New Imperial History*.

38. Vialles, *Animal to Edible*.

39. Pachirat, *Every Twelve Seconds*. See also Lee, *Meat, Modernity*; Shukin, *Animal Capital*.

40. See Gabriel N. Rosenberg for an analysis of this point in American hog production. Rosenberg, "Race Suicide among the Hogs."

41. Colley, *Britons*; Shapin, "'You Are What You Eat'"; Rogers, *Beef and Liberty*.

42. They did travel in great numbers to South America, but this is largely beyond the purview of this study.

43. Dunlap, *Nature and the English Diaspora*; Belich, *Replenishing the Earth*.

44. Carter, *His Majesty's Spanish Flock*.

45. Heath-Agnew, *History of Hereford*. See also Ritvo, *Animal Estate*, for pedigree Shorthorn breeding in the nineteenth century.

46. Collins, "Food Supplies and Food Policy."

47. Peden, "Pastoralism and the Transformation of the Open Grasslands."

48. Recent work by James Belich and Frances Steel examine the importance of steam transport to Australasian colonization, Steel for the role it played in stimulating the development of a regional identity for "Oceania," Belich, for its reinforcement of wider imperial ties. Steel, *Oceania under Steam*; Belich, *Replenishing the Earth*.

49. The phrase is borrowed from *The Corriedale, New Zealand's Own Breed: History of Development*. See also Holford, *Contribution to the Sheep World*.

50. This chapter is based in part on oral history interviews conducted with Traditional Hereford Breeders in England in 2009–10, as well as on archival research.

51. For post–World War II British immigration, particularly from the colonies and former colonies, see the following: Hansen, *Citizenship and Immigration in Post-War Britain*; Karatani, *Defining British Citizenship*; Ryan and Webster, *Gendering Migration*.

52. For some of the cultural effects of colonial and postcolonial immigration, see Sauerberg, *Intercultural Voices*; and Dawson, *Mongrel Nation*.

53. "Sir Alfred Haslam, KT., J.P.: A Sketch of His Career," in *The Queen's State Visit to Derby May 21st, 1891*, 144. Derbyshire Record Office, D1333 Z/Z 8.

Chapter One

1. Brown, *Sheep Farming*, 39.
2. Ibid., 67.
3. Ibid., 29.
4. Ibid.
5. For definitions of "breed," and for the challenge of defining breeds, see Ritvo, *Platypus and Mermaid*, esp. 78–81. See also the work of Juliet Clutton-Brock, especially *Natural History of Domesticated Mammals*; Darwin, *On the Origin of Species*, chap. 1; Darwin, *Variation of Animals*, vol. 1.
6. Coventry, *Remarks on Live Stock*, 36; Ritvo, *Platypus and Mermaid*, 81.
7. Sebright, *Art of Improving*, 12–13.
8. Lawrence, *General Treatise*, 31.
9. Darwin, *Variation of Animals*, 1:2.
10. Ibid., 3.
11. Jonsson, *Enlightenment's Frontier*; Zilberstein, *Temperate Empire*.
12. Prince, "Changing Landscape," 44–46; Turner, *Enclosure in Britain*, 28–32, 64–67, 82; Wild, *Village England*, 22–44. What constituted "improvement" varied from region to region, including, for example, land drainage as well as enclosure in Lincolnshire. See Fussell, "Four Centuries of Lincolnshire Farming," 9–10.
13. Morgan and Mingay, "Root Crops," 296–304, esp. 299–300.
14. John Sinclair, *Observations*, iv. See also Prince, "Changing Landscape," 30–41.
15. Harriet Ritvo explores the ability to formulate a "genetic template" in purebred livestock in the late-eighteenth-century livestock breeding in "Possessing Mother Nature."
16. Arthur Young's 1769 *A Six Weeks' Tour through the Southern Counties of England and Wales* is generally held to be the first of this genre, and the inspiration for the series of "general views" of the various counties of the United Kingdom subsequently commissioned by the Board of Agriculture. See Young, *Six Weeks' Tour*. Fredrik Jonsson has recently argued that the form of natural history expertise produced by this kind of reportage was a significant factor in the development of "civil cameralism" in the Scottish Enlightenment, and ensuing debates over population ecology, demography, and the future of Scotland. Jonsson, *Enlightenment's Frontier*, esp. 43–68.
17. Gooch, *Cambridge*, 266.
18. Worgan, *Cornwall*, 137.
19. Ibid.
20. Ibid.
21. Pearce, *Berkshire*, 46. See also Wood, "Sheep Breeders' View," 230.
22. Contemporary theories of epigenetics are the most important caveat to this claim.
23. Sinclair, *Observations*, ii.
24. Brown, *Sheep Farming*, 80.
25. Sinclair, *Observations*, ii.
26. Brown, *Sheep Farming*, 78.
27. Ibid., 2, 1.
28. Ibid., 78.

29. Marshall, *Rural Economy of Yorkshire*, 2:154.

30. "On the Improvement of the Highlands," *Scots Magazine*, December 1774, 643.

31. Ibid., 644.

32. Ibid.

33. Importantly, the ascent of the gene in the twentieth century has not wholly erased the significance of tacit or embodied knowledge to selective breeding. See Grasseni, *Skilled Visions*; Theunissen, "Breeding without Mendelism."

34. Brown, *Sheep Farming*, 120.

35. Ibid., 34.

36. "Experience," *Livestock Journal and Fancier's Gazette*, 63 (20 August 1875): 399.

37. Ibid., 400.

38. Ibid.; Wood, "Sheep Breeders' View," 229.

39. John Little, *Practical Observations*, 118, i.

40. Ibid., ii.

41. Ibid., ii–iii.

42. Ibid., iv.

43. Ibid., 118.

44. Derry, *Masterminding Nature*. The monastic sheep breeders in Silesia who selected according to Mendel's theory inheritance for decades before "Mendelian genetics" were "rediscovered" were the major exception to this. See Wood and Orel, *Genetic Prehistory*; Wood, "Sheep Breeders' View."

45. Hunt, *Memoirs*, 21.

46. "Breeding for Fertility," *New Zealand Farmer* 22, no. 9 (September 1902): 468.

47. While breeders sought tangible profits in the form of their stock's progeny, early geneticists were preoccupied with determining the principles underlying processes of inheritance. Not until the 1920s did these disparate aims converge with the science of livestock husbandry. Derry, *Masterminding Nature*.

48. Lawrence, *General Treatise*, 28. Müller-Wille and Rheinberger, "Heredity"; Wood, "Sheep Breeders' View"; Wood and Orel, *Genetic Prehistory*.

49. Lawrence, *General Treatise*, 28.

50. Ibid. See also Russell, *Like Engend'ring Like*.

51. Quoted in Hunt, *Memoirs*, 17.

52. Derry identifies the "progeny test" as a major feature of eighteenth and nineteenth-century livestock breeding, and as one of the dividing lines between the sciences of heredity and the practices of breeding. Derry, *Masterminding Nature*.

53. Coventry, *Remarks on Live Stock*, 5.

54. Ibid., 6.

55. Sebright, *Art of Improving*, 6.

56. Ibid., 8.

57. Ibid., 5.

58. Ibid., 6–7.

59. Coventry, *Remarks on Live Stock*, 36–37.

60. Agricolanus, "Directions for Raising and Managing Sheep," *American Museum, or, Repository of Ancient & Modern Fugitive Pieces, &c. Prose & Poetical* 2, no. 3 (September 1787): 295.

61. Agricola, "Letters on the Improvement of the Highlands of Scotland: The Influence of Climate upon the Quality of Wool," *Scots Magazine*, October 1774, 529.

62. Ibid.

63. Lawrence, *General Observations*, 306.

64. "Improvement of the Highlands," 646.

65. Ibid.

66. Ibid., 645.

67. "Natural History of the Sheep," *Monthly Miscellany* 2, no. 12 (December 1774): 303.

68. Chaplin, "Creoles in British America"; Harrison, *Climates & Constitutions*; Shapin, "'You Are What You Eat.'"

69. Chaplin, "Creoles in British America"; Chaplin, *Subject Matter*; Kupperman, "Fear of Hot Climates." For colonial racial vulnerability in later centuries, see Warwick Anderson, *Cultivation of Whiteness*; Warwick Anderson, "Disease, Race, and Empire"; Kennedy, "Perils of the Midday Sun."

70. For the analogy, see Ritvo, *Platypus and Mermaid*, 75–81, 121–27; Pawley, "Point of Perfection."

71. Lawrence, *General Observations*, 307.

72. Ibid. Now largely out of usage, the term "shangalla" referred to peoples of neither Ethiopian nor Arab descent in northeast Africa. See, for example, Koettlitz, *Journey Through Somaliland*.

73. Rudge, *Gloucester*, 305, 307–9. The Cotswolds are a region in England that encompass portions of Worcestershire, Gloucestershire, Wiltshire, Oxfordshire, Warwickshire, and Somersetshire.

74. Worgan, *Cornwall*, 148.

75. Wood, "Sheep Breeders' View," 232–36; Ritvo, "Possessing Mother Nature."

76. Somerville, *Facts and Observations*, 3.

77. Ibid., 2.

78. Brown, *Sheep Farming*, 29, 27.

79. Archer and Sinclair, *Domestic Breeds and Their Treatment*, 12.

80. Brown, *Sheep Farming*, 29.

81. Ibid.

82. Ibid.

83. The other improved breed of sheep with a significant impact at this time was the Southdown breed.

84. Ritvo, "Possessing Mother Nature," 416.

85. Ibid., 418.

86. Rudge, *Gloucester*, 307.

87. Worgan, *Cornwall*, 148.

88. Ibid., 305; for the extinction of the Norfolk Horn breed of sheep in the mid-nineteenth century, see Low, *Breeds of the Domestic Animals*, 116.

89. Brown, *Sheep Farming*, 28.

90. Trow-Smith, *History of British Livestock*, vol. 2.

91. Langford, "Eighteenth Century," 440–47.

92. For an example of a work that locates this at the center of British identity, see Rogers, *Beef and Liberty*. See also Ritvo, *Platypus and Mermaid*, 200.

93. Brown, *Sheep Farming*, 31.

94. James Vernon discusses the inverse of this—hunger as political critique—in *Hunger*. For the impact of demography in world history, and particularly the effect of Europe's population explosion in the early modern period, see Pomeranz, *Great Divergence*. See also Otter, "Civilizing Slaughter."

95. Hunt, *Memoirs*, 22–23.

96. Ibid.

97. Sinclair, *Observations*, xviii.

98. Ibid.

99. Rudge, *Gloucester*, 305–6.

100. Ibid.

101. Shapin, "'You Are What You Eat.'"

102. "Our Meat-Supply," *Chambers's Journal* 257 (28 November 1868): 760.

103. Quoted in "Typical Differences in English and French Beef Cattle," *New Zealand Farmer* 21, no. 10 (October 1901): 444.

104. "Imported Cattle and Disease," *Livestock Journal* 2 (27 August 1875): 424.

105. Brown, *Sheep Farming*, 24.

106. Ibid., 115.

107. Ibid.

108. Ibid., 28.

109. Worgan, *Cornwall*, 149.

110. Brown, *Sheep Farming*, 99.

111. Ibid., 116.

112. Sinclair, *Observations*, v–vi. Original spelling preserved.

113. Brown, *Sheep Farming*, 38.

114. Ibid., 29.

115. Ibid., 116.

116. Quoted in Archer and Sinclair, *Domestic Breeds and Their Treatment*, 13.

117. F. Boys, "Agricultural Minutes, Taken during a Ride through the Counties of Kent, Essex, Suffolk, Norfolk, Cambridge, Rutland, Leicester, Northampton, Buckingham, Bedford, Hertford, Middlesex, Berks, and Surry, in 1792," *Annals of Agriculture* 19 (1793): 120.

118. See chapter 4.

119. Gooch, *Cambridge*, 266.

120. The extreme pitch of the topography in some parts of New Zealand necessitated breeding for "well-sprung" hocks—the joints of the hind legs—in cattle. Oral history interview with Philip Barnett, Akitoa, New Zealand, 24 June 2010. See also Peden, "Pastoralism and the Transformation."

121. Brown, *Sheep Farming*, 79.

122. Lawrence Alderson, "Conserving the Cattle of Britain," *Ark* no. 4, May 1977, 157.

Chapter Two

1. Arthur Young, "Don Merino," *Annals of Agriculture* 17 (1792): 531.

2. C. Mordaunt, "Lancashire Improvements," *Annals of Agriculture* 19 (1793): 253.

3. Joseph Banks, "A Project for Extending the Breed of Fine-Wooled Spanish Sheep, Now in the Possession of His Majesty, into All Parts of Great Britain, Where the Growth of Fine Clothing Wools Is Found to Be Profitable," *European Magazine*, 1800, 175.

4. George Tollet, "Merino Sheep," *Annals of Agriculture* 44, no. 256 (1806): 9.

5. Ibid., 10.

6. Practicus, "Remarks on the Duke of Bedford's Discontinuing His Premiums to the New Leicester and Southdown Breed of Sheep, and on Lord Somerville's and Dr. Parry's Encouragement of the Spanish Breed," *Agricultural Magazine, or, Farmers' Monthly Journal of Husbandry and Rural Affairs* 6, no. 35 (June 1802): 434.

7. C. H. Parry, "Dr. Parry, in Answer to Practicus, on the Breed of Sheep," *Agricultural Magazine, or, Farmers' Monthly Journal* 7, no. 36 (July 1802): 8, 9. Cf. Parry, *Practicability and Advantage*.

8. John Hunt, "Perfections and Superiority of the Leicestershire Breed," *Agricultural Magazine* 3, no. 14 (August 1808): 88, 90; John Hunt, "On the Imperfections and Inferiority of the Merino Sheep; and the Impropriety of Introducing Them into This Country, in Answer to Mr. Thompson," *Agricultural Magazine, or, Farmers' Monthly Journal* 4, no. 19 (January 1809): 57; Cultivator Middlesexiensis, "On the New Leicester and the Merino Sheep, in Answer to Mr. Hunt," *Agricultural Magazine, or, Farmers' Monthly Journal* 3, no. 15 (September 1808): 188; Benjamin Thompson, "Refutation of Mr. Hunt's Absurdities," *Agricultural Magazine, or, Farmers' Monthly Journal* 3, no. 18 (December 1808): 360; Benjamin Thompson, "The Merino Cause—Description of His Majesty's Spanish Sheep—and Final Reply to the Dishley Quack," *Agricultural Magazine, or, Farmers' Monthly Journal* 4, no. 21 (March 1809): 160.

9. Practicus, "Remarks on the Duke of Bedford," 434.

10. Carter, *His Majesty's Spanish Flock*, 3–4, 8, 11; Trow-Smith, *History of British Livestock*, 1:39.

11. Trow-Smith, *History of British Livestock*, 1:39.

12. Carter, *His Majesty's Spanish Flock*, 3, 44.

13. See Ritvo, "Possessing Mother Nature"; Franklin, *Dolly Mixtures*, chap. 2, esp. 52–54.

14. Rudge, *Gloucestershire*, 312. See also Ritvo, "Possessing Mother Nature."

15. Carter, *His Majesty's Spanish Flock*, 6–8.

16. First published as "Account of the Sheep and Sheep-Walks of Spain," *Scots Magazine*, 1764, 361–68. Abridged and reprinted as "The Method of Managing the Royal Flocks of Sheep in Spain," *Columbian Magazine; or, Monthly Miscellany: Containing a View of the History, Literature, Manners & Characters of the Year*, 1789, 475–79. But this is hardly a comprehensive list. Citations for this chapter are drawn from a three-part series of the full letter, "Account of the Sheep and Sheep-Walks of Spain, in a Letter from a Gentleman in Spain to Mr. Peter Collinson, F.R.S.," *New York Magazine*, August 1790, 454–57; ibid., September 1790, 518–21; and ibid., October 1790, 567–71.

17. "Sheep and Sheep-Walks of Spain," 2:519.

18. Ibid.
19. Ibid., 3:570.
20. For example, de Lasteyrie, *Introduction of Merino Sheep*. First published in French in 1802. See also "Memoir on the Management of Sheep, at Cauterrets; A Department in the Basses Pyrenees. From the French of M. Jenow," *Annals of Agriculture* 17 (1792).
21. "Sheep and Sheep-Walks of Spain," 1:455.
22. Culley, *Observations on Live Stock* (1807), 237.
23. Carter, *His Majesty's Spanish Flock*, 49–53.
24. Culley, *Observations on Live Stock* (1807), 237.
25. Carter, *His Majesty's Spanish Flock*, 54–59; Banks, *Circumstances Relative to Merino Sheep*, 7.
26. Banks, *Project for Extending the Breed*, 1. In return, George III sent "eight fine English coach horses" to the Marchioness. Ibid.
27. Banks, *Circumstances Relative to Merino Sheep*, 8.
28. Ibid., 10.
29. Joseph Banks, "A Report of the State of His Majesty's Flock of Fine-Wooled Spanish Sheep, during the Years 1800 and 1801; With Some Account of the Progress That Has Been Made towards the Introduction of That Valuable Breed into Those Parts of the United Kingdom Where Fine Cloathing Wools Are Grown with Advantage," *Annals of Agriculture* 40, no. 233 (1803): 357.
30. Banks, *Circumstances Relative to Merino Sheep*, 8.
31. "Sale of Part of His Majesty's Flock of Spanish Sheep," *Agricultural Magazine, or, Farmers' Monthly Journal* 11, no. 61 (August 1804): 145.
32. Ibid., 146.
33. Ibid., 145.
34. Ibid., 147.
35. Ibid., 144.
36. "The King's Annual Sale of Sheep," *Agricultural Magazine, or, Farmers' Monthly Journal* 13, no. 73 (August 1805): 132.
37. Ibid.
38. Ibid., 134.
39. At subsequent auctions, prices realized only continued to rise, peaking at £74 for a ram in 1808. Banks, *Circumstances Relative to Merino Sheep*, 9.
40. Banks, "Report 1800 and 1801," 356.
41. Ibid.
42. Ibid., 354.
43. Ibid., 355.
44. Carter, *His Majesty's Spanish Flock*, 426.
45. Figures from Mitchell, *Abstract of Statistics*, 191. See app. A.
46. Somerville, *Somerville's Address to the Board*, 3.
47. Broers, *Europe under Napoleon*, 96.
48. Bucke, "Report," 17; Banks, "Address to the Members," 6.
49. Montalivet, "Report of the Minister of the Interior," 48; Benjamin Thompson, "Preface to First Report," iii.

50. Bucke, "Report," 18.

51. Benjamin Thompson, "Appendix: Letter from Thompson to Banks," 138–39.

52. George Hall, "Growth and Management of Merino Wool," 45.

53. Benjamin Thompson, "Appendix: Letter from Thompson to Banks," 138.

54. George Hall, "Growth and Management of Merino Wool," 42; Banks, *Circumstances Relative to Merino Sheep*, 5.

55. Benjamin Thompson, "Appendix: Letter from Thompson to Banks," 159.

56. Somerville, *Somerville's Address to the Board*, 3.

57. Rudge, *Gloucestershire*, 313; Banks, "Address to the Members," 5; Bucke, "Report," 20.

58. Benjamin Thompson, "Merino Cause," 160.

59. John Hunt, "Perfections and Superiority," 91; John Hunt, "Commercial Philosophy, or an Address to Mr. Robert Bakewell of Wakefield in Answer to His Observations on the Influence of Soil and Climate upon Wool," *Agricultural Magazine, or, Farmers' Monthly Journal* 3, no. 15 (September 1808): 185.

60. Pastorius, "On Spanish Sheep," *Agricultural Magazine, or, Farmers' Monthly Journal* 11, no. 63 (October 1804): 240. Italics in original.

61. Ibid.

62. Jonsson, *Enlightenment's Frontier*, esp. chap. 1.

63. John Hunt, "On the Merino Question: The Critic Unmasked, or Truth without Disguise, in Answer to Cultivator Middlesexiensis," *Agricultural Magazine, or, Farmers' Monthly Journal* 3, no. 17 (November 1808): 313.

64. Pastorius, "Spanish Sheep," 242–43.

65. Benjamin Thompson, "On Merino, New Leicester Sheep, &c. in Answer to Mr. Hunt," *Agricultural Magazine, or, Farmers' Monthly Journal* 3, no. 16 (October 1808): 223.

66. Sebright, *Art of Improving*, 3.

67. Newnham, quoted in Bucke, "Report," 73; Benjamin Thompson, "Appendix: Letter from Thompson to Banks," 146.

68. "On the Improvement of the Highlands," *Scots Magazine*, 1774, 645.

69. Sebright, *Art of Improving*, 18.

70. John Hunt, "Perfections and Superiority," 92.

71. Ibid., 84.

72. Ibid., 86–87.

73. "Improvement of the Highlands," 644.

74. Cultivator Middlesexiensis, "New Leicester and Merino Sheep," 189.

75. "Improvement of the Highlands," 646.

76. Ibid., 644. The major exception to this is overseas colonization (discussed in chapters 4 and 5) and the provisioning of ships (the subject of an article in production).

77. Sebright, *Art of Improving*, 20.

78. Pearce, *Berkshire*, 46. See also Wood, "Sheep Breeders' View," 230.

79. See also chapter 1.

80. F. H. Clay, "State of the Merino Improvement in Sherwood Forest Notts," *Agricultural Magazine, or, Farmers' Monthly Journal* 3, no. 18 (December 1808): 357.

81. Newnham, quoted in Bucke, "Report," 73.
82. Ibid., 72.
83. Unattributed, quoted in Bucke, "Report," 64.
84. Benjamin Thompson, "Merino Cause," 155.
85. Somerville, *Facts and Observations*, 15.
86. Sebright, *Art of Improving*, 20; Bucke, "Report," 8.
87. Benjamin Thompson, "Appendix: Letter from Thompson to Banks," 119.
88. Bucke, "Report," 9.
89. John Wright, "On Merino and New Leicester Sheep, in Answer to Mr. Hunt," *Agricultural Magazine, or, Farmers' Monthly Journal* 5, no. 25 (July 1809): 13.
90. Ibid., 9.
91. Agricola Northumbriensis, "On Merino Sheep, with Miscellaneous Observations," *Agricultural Magazine, or, Farmers' Monthly Journal* 5, no. 28 (October 1809): 243; Bucke, "Report," 71.
92. Agricola Northumbriensis, "On Merino Sheep," 243.
93. George Hall, "Growth and Management of Merino Wool," 49.
94. Ibid., 46.
95. Charles Hunt, *Merino and Anglo-Merino Breeds of Sheep*, 15.
96. John Hunt, "Perfections and Superiority," 90.
97. Sebright, *Art of Improving*, 22.
98. Somerville, *Facts and Observations*, 40–41.
99. Banks, *Project for Extending the Breed*, 6.
100. John Hunt, "Perfections and Superiority," 88.
101. Somerville, *Facts and Observations*, 3.
102. Parry, *Practicability and Advantage*, 42.
103. George Hall, "Growth and Management of Merino Wool," 52.
104. Somerville, *Facts and Observations*, 19.
105. Ibid.
106. Benjamin Thompson, "Appendix: Letter from Thompson to Banks," 119.
107. Bucke, "Quality of the Mutton," 13.
108. Lord Sheffield, quoted ibid., 16.
109. Ritvo, *Platypus and Mermaid*, 196–97; Banks, *Circumstances Relative to Merino Sheep*, 4.
110. Charles Hunt, *Merino and Anglo-Merino Breeds of Sheep*, 72.
111. Somerville, *Facts and Observations*, 15.
112. Ibid., 15–16.
113. Cultivator Middlesexiensis, "New Leicester and Merino Sheep," 192.
114. Parry, *Practicability and Advantage*, 36.
115. Ibid., 42.
116. John Hunt, "Perfections and Superiority," 88.
117. Benjamin Thompson, "Appendix: Letter from Thompson to Banks," 121.
118. Cultivator Middlesexiensis, "New Leicester and Merino Sheep," 192, 193.
119. Quoted in Charles Hunt, *Merino and Anglo-Merino Breeds of Sheep*, 117–18.
120. Benjamin Thompson, "Refutation of Mr. Hunt's Absurdities," 360.

121. Benjamin Thompson, "Successful Experiment of a Merino-Shetland Cross-Sheep-Sheering," *Agricultural Magazine, or, Farmers' Monthly Journal* 4, no. 24 (June 1809): 359.

122. John Wright, "A Comparative View of the New Leicester and Half-Bred Merino Sheep," *Agricultural Magazine, or, Farmers' Monthly Journal* 3, no. 18 (December 1808): 364.

123. "Extract from a Report on Lord Somerville's Show," quoted ibid., 8.

124. "Extract from a Report on Lord Somerville's Show," quoted in Bucke, "Report," 7.

125. Ibid.

126. Mitchell, *Abstract of Statistics*, 192.

Chapter Three

1. Marshall, *Rural Economy of Gloucestershire*, 187, 188.

2. Ibid., 187. See also Duncumb, *General View of Hereford*.

3. Marshall, *Rural Economy of Gloucestershire*, 192.

4. Ibid., 193.

5. Duckham, "Breeding and Management," 4. Paper originally given at the Breconshire Chamber of Agriculture, 2 January 1869.

6. Youatt, *Cattle*, 32; Marshall, *Rural Economy of Gloucestershire*, 193.

7. Youatt, *Cattle*, 31; quoted from the *Hereford Times* in "Hereford Cattle," *Maitland Mercury & Hunter River General Advertiser* (New South Wales), 3 October 1885, supp.: 21; Duckham, "Breeding and Management," 5.

8. Marshall, *Rural Economy of Gloucestershire*, 192. Juliet Clutton-Brock and Steven Hall call the Hereford "probably the most famous county breed of cattle." Clutton-Brock and Hall, *British Farm Livestock*, 76.

9. Youatt, *Complete Grazier*, 12. See also chapter 1.

10. Culley, *Observations on Live Stock* (1807), v.

11. Youatt, *Cattle*, 9.

12. Ibid., 11, 9.

13. Ibid., 24.

14. "Hereford Cattle," *Livestock Journal and Fancier's Gazette* 2 (12 November 1875): 688.

15. "Cattle of the Various Breeds as Beef Producers," *Farmer's Magazine* 55 (February 1879): 99.

16. Ibid.

17. Youatt, *Cattle*, 11. Bridgewater, at the mouth of the river Taw, is a mere forty miles west of where the river Parrett meets the Bristol Channel.

18. "Among the Herefords: Mr. Boughton-Knight's Herd at Leinthall," *Livestock Journal and Fancier's Gazette* 21 (2 April 1885): 327.

19. John Speed, *England, Wales, and Scotland Described*, 1627. Quoted in MacDonald and Sinclair, *History of Hereford Cattle*, 1.

20. George Garrard, *A Description of the Different Oxen Common in the British Isles*, 1800. Quoted in MacDonald and Sinclair, *History of Hereford Cattle*, 7.

21. Duckham, "History, Progress, and Comparative Merits," 32. Paper originally given at the Royal Agricultural College at Cirencester, 4 December 1863.
22. Youatt, *Cattle*, 32.
23. "The Hereford Cattle Outlook," *Launceston Examiner*, 6 April 1881, supp.: 2.
24. Duckham, "History, Progress, and Comparative Merits," 4.
25. Thomas Andrew Knight, "Account of the Herefordshire Breeds of Sheep, Cattle, Horses, and Hogs," *Commercial and Agricultural Magazine* 7, no. 40 (November 1802): 334; Duckham, "History, Progress, and Comparative Merits," 8.
26. Duckham, "History, Progress, and Comparative Merits," 8.
27. William Cronon explores the relationship of production, consumption, and extraction between a metropolis and its hinterlands in the American context, James Belich in the global context. Cronon, *Nature's Metropolis*; Belich, *Replenishing the Earth*. For a discussion of how livestock reached London in the nineteenth century, see Metcalfe, *Meat, Commerce and the City*, 17–21; Trow-Smith, *History of British Livestock*, 2:3–10; 172–73; 226–28.
28. Trow-Smith, *History of British Livestock*, 2:45–46. See also Ritvo, *Animal Estate*, 47.
29. Youatt, *Cattle*, 4.
30. Fowler, *Records of Old Times*, 92.
31. See Freeman, *Mutton and Oysters*, 178–210, on changes to menus and dining habits in the nineteenth century.
32. Ritvo, *Animal Estate*, chap. 1, esp. 56.
33. "Extracts from Minutes of the Smithfield Club," 27; Duckham, "History, Progress, and Comparative Merits," 7.
34. "On the Late Cattle Show, with Remarks," *Agricultural Magazine, or, Farmers' Monthly Journal*, no. 1, new ser., January 1813, 31.
35. A number of agricultural worthies (including Westcar) together formed the Smithfield Club for the purpose, according to Powell, of "bringing out . . . the principle of early maturity." Powell, *History of the Smithfield Club*, 1.
36. Culley, *Observations on Live Stock* (1807), 46.
37. "On the Late Cattle Show," 32.
38. Youatt, *Complete Grazier*, 9.
39. Duncumb, *General View of Hereford*, 116.
40. Youatt, *Complete Grazier*, 9.
41. Clutton-Brock and Hall, *British Farm Livestock*, 76.
42. "Hereford Cattle," *Livestock Journal* 18 (14 June 1882). Repr. in *Maitland Mercury*, 26 May 1883, 6.
43. Duckham, "Breeding and Management," 5.
44. Duncumb, *General View of Hereford*, 118.
45. "Imported Hereford Cattle," *Maitland Mercury*, 22 March 1879, 6.
46. Youatt, *Complete Grazier*, 9.
47. Duncumb, *General View Hereford*, 119.
48. Thomas Duckham, quoted in "The Hereford Breed of Cattle," *Mercury* (Hobart, Tasmania), 3 May 1872, 3.
49. Duncumb, *General View of Hereford*, 119.
50. "Hereford Cattle," *Maitland Mercury*, 3 October 1885, supp.: 21.

51. Ritvo, "Possessing Mother Nature."

52. T. Weston, "General Remarks on the Shew of Fat Cattle in Smithfield," *Commercial and Agricultural Magazine* 5, no. 29 (December 1801): 383.

53. "Hereford Cattle," repr. from the *Pacific Rural Press* in *South Australian Register*, 14 December 1877, 9.

54. Youatt, *Cattle*, 68, 67.

55. "Hereford Cattle," *South Australian Register*, 1877, 9.

56. "Hereford Cattle," *Brisbane Courier*, 13 December 1882, 3.

57. Knight, "Account of Herefordshire Breeds," 332.

58. "Hereford Breed of Cattle," 3.

59. Ibid.

60. "Remarks and Observations on Different Kinds of Cattle, Continued from Our Last," *Agricultural Magazine, or, Farmers' Monthly Journal* 7 (December 1810): 390. Richard Parkinson remarked that he did not believe there was "a single cow to be found in the possession of any cow-keeper in London of the Hereford breed." Parkinson, *Breeding and Management of Live Stock*, 111.

61. As William Youatt remarked in the preface to *Cattle: Their Breeds, Management, and Diseases* (1834), so strong did feelings of partiality run among breeders that "although there is some excellence peculiar to each breed, there is none exempt from defect, and the honest statement of this defect will not satisfy the partisan of any one breed." Youatt, *Cattle*, iii.

62. "Important to Dairymen: Herefords and Short-horns," *Farmer's Magazine* 9 (May 1844): 555.

63. Quoted from *The Field* in "Hereford Breed of Cattle," 3.

64. "Cattle as Beef Producers," 100.

65. Youatt, *Complete Grazier*, 9.

66. "Cattle as Beef Producers," 100.

67. Duckham, "Breeding and Management," 7.

68. "The Humble Petition of 500,000 Frugally Disposed Housekeepers, Resident in the United Kingdoms of England, Scotland, and Ireland," *Commercial and Agricultural Magazine* 3, no. 17 (December 1800): 404.

69. Ibid.; T. Weston, "Answer to the Petition of 500,000 Housekeepers," *Commercial and Agricultural Magazine* 4, no. 18 (January 1801): 6.

70. "Humble Petition," 404.

71. Ibid.

72. Weston, "Answer to the Petition," 6.

73. Ibid., 7.

74. Ibid., 6.

75. Ibid.

76. Ibid., 8.

77. Quoted in Weston, "Shew of Fat Cattle," 383.

78. Ritvo, *Animal Estate*, 72–74.

79. As proof, Weston cited a recent decision to give preference to one of Westcar's oxen—of a larger size but less fat—over a fatter animal. Weston, "Shew of Fat Cattle," 383.

80. "Proceedings of Agricultural Societies: Smithfield Club," *Agricultural Magazine, or, Farmers' Monthly Journal* 13, no. 77 (December 1805): 431.
81. "Remarks and Observations, Continued," 395.
82. Ibid., 396.
83. See chapter 1 for a discussion of this in reference to sheep, especially the Dishley breed. See also Ritvo, *Animal Estate*, 17.
84. MacDonald and Sinclair, *History of Hereford Cattle*, 6.
85. "Remarks and Observations, Continued," 396.
86. Powell, *Smithfield Club*, 2, 1.
87. Duncumb, *General View Hereford*, 177.
88. Weston, "Shew of Fat Cattle," 383.
89. Ritvo, "Possessing Mother Nature."
90. Coates, *Short-Horned Herd-Book*; Eyton, *Herd Book of Hereford Cattle*, vol. 1.
91. "Hereford Cattle," *Illustrated Sydney News*, 19 March 1881, 15.
92. Duckham, "History, Progress, and Comparative Merits," 32.
93. George Culley, *Observations on Live Stock* (1786), 21. Also quoted in J. H. Campbell, "On the Breeds of Cattle and Sheep," *Annals of Agriculture* 16 (1790): 227.
94. Campbell, "Breeds of Cattle," 226.
95. Ibid.
96. Ibid. For his part, upon further consideration, Culley replied to Campbell that he was willing to revise his position and "to suppose they may be an original breed," and he promised to correct his mistake in future editions. See George Culley, "On Cattle," *Annals of Agriculture* 16 (1790): 181. Culley made good on that promise, eliminating the offending remarks entirely from his description of the "Herefordshire Cattle." Culley, *Observations on Live Stock* (1807), 52–53.
97. Q., "Remarks on the Late Cattle Show," *Agricultural Magazine, or, Farmers' Monthly Journal* 8 (January 1811): 15.
98. Ibid.
99. Ibid., 14.
100. Youatt, *Cattle*, 9.
101. Weston, "Shew of Fat Cattle," 383.
102. "Remarks and Observations on Different Sorts of Cattle," *Agricultural Magazine, or, Farmers' Monthly Journal* 7 (November 1810): 326.
103. T. S., "On the Choice and Management of Dairy Stock, with a Few Observations on the Best Methods of Rearing Calves," *Agricultural Magazine, or, Farmers' Monthly Journal* 3 (July 1808): 7.
104. "Hereford Cattle," *Livestock Journal* 2 (12 November 1875): 688.
105. Youatt, *Cattle*, 9.
106. That is, without horns. Ibid., 188.
107. Youatt, *Complete Grazier*, 11.
108. Breeds displaying markings of this sort are today called "belted" cattle. The "barrel" is the torso of an animal. Youatt, *Cattle*, 28.
109. Ibid., 9.
110. The great diversity of domesticates, bovine or otherwise, has at one time or another suggested to observers multiple moments of domestication for a given

species. Current theory holds, and is supported by genetic evidence, that each species was domesticated only once.

111. Culley, *Observations on Live Stock* (1807), 55.

112. Youatt, *Cattle*, 9. Though Bakewell is credited with "improving" the Longhorn type native to Lancashire in the late eighteenth century, his methods were less effective upon cattle than sheep, and the Improved Longhorn was never as widely adopted, or as loudly applauded, as its contemporary, the Improved Shorthorn. Ritvo, "Possessing Mother Nature"; Trow-Smith, *British Livestock Husbandry*, 83–89. In the mid-1980s, the Longhorn Cattle Society of England continued to tout the English Longhorn as "Britain's oldest Beef Breed." Quoted in Worcester, *Texas Longhorn*, 14.

113. Youatt, *Cattle*, 9.

114. Marshall, *West of England*, 236.

115. Ibid.

116. Youatt, *Cattle*, 9.

117. Longhorns, according to Youatt, were "evidently of Irish extraction," and Shorthorns of even more "foreign" extraction. Ibid.

118. Ritvo, "Race, Breed, and Myths of Origin," 140–41. Ritvo notes that nineteenth-century antiquarians erroneously connected these so-called wild park cattle to a pre-Roman type found in the south of England and the midlands. Ritvo, *Animal Estate*, 46, 300n4. Bovines were originally domesticated in western Asia and southeast Europe, and in northern Europe, they "probably resembled quite closely the modern Dexter breed." Juliet Clutton-Brock, *Natural History of Domesticated Mammals*, 68.

119. Marshall, *West of England*, 236.

120. Ritvo, "Race, Breed, and Myths of Origin," esp. 148.

121. Youatt, *Cattle*, 9, 10.

122. Ibid., 10.

123. Ibid. In this, he followed Marshall, who wrote that "their color apart, they nearly resemble the wild cattle which are still preserved in Chillingham Park, in Northumberland." Marshall, *West of England*, 236.

124. Youatt, *Cattle*, 10.

125. Ibid., 32.

126. Campbell, "Breeds of Cattle," 226.

127. Duckham, "History, Progress, and Comparative Merits," 10.

128. Ibid., 13, 15.

129. Thomas Andrew Knight was president of the London Horticultural Society during his lifetime, and also the author of a number of scientific papers on plant and animal breeding, including "Experiments of the Fecundation of Vegetables" and "Hereditary Instinctive Propensities."

130. "Among the Herefords: Boughton-Knight's Herd," 327.

131. Coates, *Short-Horned Herd-Book*.

132. Youatt, *Cattle*, 226; "Herefords in Westmeath," repr. from *The Irish Farmers' Gazette* in the *Livestock Journal* 2 (3 September 1875): 450.

133. Duckham, "History, Progress, and Comparative Merits," 9.

134. Ibid.

135. Duncumb, *General View Hereford*, 116.

136. Fowler, *Records of Old Times*, 96. He described himself as "for many years an ardent admirer and somewhat successful breeder of Shorthorns." Ibid., 95.

137. Ibid., 96. Duncumb described the Hereford fair in similar terms: "The shew of oxen in thriving condition at the Michaelmas fair in Hereford, cannot be exceeded by any similar annual collection in England." Duncumb, *General View Hereford*, 116. The Devons were considered closely related to the Hereford, although more frisky and not as easily fattened.

138. Welles, *Remarks and Suggestions*, 6, 7, 11. The chine refers to the spine and back of an animal.

139. Quoted from the *Hereford Times* in "Hereford Cattle," *Maitland Mercury*, 3 October 1885, supp.: 3.

140. In the 1790s, the characteristic color of "the true breed" was "a middle red [with] a 'bald face,'" according to MacDonald and Sinclair, and in 1802 Knight described the "Herefordshire colour" as "a deep red, with a white face." MacDonald and Sinclair, *Hereford Cattle*, 5; Knight, "Herefordshire Breeds," 332.

141. Thomas Andrew Knight, quoted in Duckham, "History, Progress, and Comparative Merits," 5. Ibid., 4, 5. Knight was careful, though, not to suggest that the Hereford breed itself was exogenous, only that "its superiority was attributed to the importation of Flemish cattle ... thus ... convey[ing] the impression that the infusion of the Flanders strain into the Hereford cattle had developed the good properties of the native breed to a greater extent than had before been attained." MacDonald and Sinclair, *Hereford Cattle*, 14.

142. Duckham, "History, Progress, and Comparative Merits," 5.

143. "Among the Herefords: Boughton-Knight's Herd," 327.

144. MacDonald and Sinclair, *Hereford Cattle*, 7.

145. Duckham, "History, Progress, and Comparative Merits," 12.

146. "Hereford Cattle," *Livestock Journal*, 1875, 689.

147. Ibid., 688.

148. Duckham, "History, Progress, and Comparative Merits," 9.

149. Clutton-Brock and Hall, *British Farm Livestock*, 77. Indeed, as Margaret Derry suggests, "That identification through public pedigree information was available for Shorthorns earlier than for other cattle breeds helped provide an important start to the breed's ultimate popularity and geographic expansion.... Possibly Shorthorns became so popular ... not because they were improved before other breeds ... but rather because of the head start provided by the breed's public herd book." Derry, *Bred for Perfection*, 6.

150. Derry, *Bred for Perfection*.

151. Darwin, *Variation under Domestication*, 1:2.

152. Ibid., 4.

153. Ibid.

154. Ibid.

155. Culley, *Observations on Live Stock* (1807), viii.

156. Ibid.

157. Culley, *Observations on Live Stock* (1807), viii–ix.

158. V., "Stock Breeding," *Livestock Journal* 21 (17 April 1885): 376.

159. Duckham, "Breeding and Management," 5.

160. Quoted in "Cattle of the Various Breeds as Beef Producers," *Farmer's Magazine* 55 (February 1879): 99.

161. Derry, *Bred for Perfection*, chap. 1; Ritvo, *Animal Estate*, 60–63.

162. "Animals were believed to be 'pure' to breed type ... when they carried public pedigrees.... Ideas about breed, the meaning of purity within breed, and the role of pedigrees in breeding became entangled in a complicated way." Derry, *Bred for Perfection*, 9.

163. Practice, "Stock Breeding," *Livestock Journal* 21 (10 April 1885): 350.

164. "Various Notes," *Farmer's Magazine* 55 (May 1879): 316.

165. "Cattle as Beef Producers," 100.

166. "An American on Breeding," *Livestock Journal* 2 (8 October 1875): 568.

167. "Various Notes," 316.

168. "Cattle as Beef Producers," 100.

169. Ibid.

170. Ibid.

171. "Remarks and Observations, Continued," 390.

172. "British Breeds of Cattle," *Livestock Journal* 21 (22 May 1885): 495.

173. "Cattle as Beef Producers," 100.

174. "Various Notes," 313.

175. "Herefords in Westmeath," 2.

176. Cosmo, "Among the Herefords: The Hampton Court Herefords," *Livestock Journal* 21 (17 April 1885): 373.

177. Ibid.

178. Arkwright served four terms as president, and five as vice president of the society between 1878 and 1898. MacDonald and Sinclair, *Hereford Cattle*, 144. Prior to 1878, the *Herd Book of Hereford Cattle* was privately operated.

179. Cosmo, "Among the Herefords," 373.

180. Ibid.

181. Ibid.

182. Thomas Campbell Eyton (1809–80) was a naturalist who specialized in ornithology. He was the author of a number of works, including *A History of the Rarer British Birds* and *A Catalogue of British Birds*. He was a friend and contemporary of Charles Darwin, with whom he exchanged a number of letters on zoology, the anatomy of birds, and Herefordshire cattle. See in particular, Darwin, "to T. C. Eyton 27 August 1856," http://www.darwinproject.ac.uk/ entry-1946; and Darwin, "to Eyton 31 August 1856," http:// www.darwinproject.ac.uk/entry-1948.

183. Duckham, "History, Progress, and Comparative Merits," 12.

184. J. H. Arkwright, Draft letter to the editor of the *Hereford Times* (July 1888), in response to "Hereford Herd Book Society," *Hereford Times*, 27 June 1888, signed Herefordian. Herefordshire Archive and Record Centre (hereafter HARC), A63/III/65/14.

185. Duckham, "History, Progress, and Comparative Merits," 12, 10. Exasperated, Eyton declared his "intention [not] to carry the Work on further unless the

breeders generally come forward to assist me more than they have done up to the present time" after publishing only two volumes. Eyton, "Preface," 2: iv. At this point, the Hereford herd book copyright passed to Thomas Duckham, who published seven volumes before handing it off to the Hereford Herd Book Society in 1878.

186. J. R. Bailey to J. H. Arkwright, n.d. 1884, HARC, A63/IV/42/33.
187. Duckham, *Eyton's Herd Book*, 3:iii.
188. Eyton, *Herd Book of Hereford Cattle*, 1:iii.
189. Duckham, *Eyton's Herd Book*, 3:iii.
190. Ibid., iv.
191. Cosmo, "Among the Herefords," 374.
192. Ibid.
193. Ibid. A remarkable number of these documents ended up in the Herefordshire Archive and Record Centre.
194. Hereford Herd Book Society, *Herd Book of Hereford*, 11:vii.
195. Joseph Russell Bailey to J. H. Arkwright, 26 April 1884, HARC, A63/IV/42/33.
196. Ibid.
197. Ibid.
198. Percy Powell to J. H. Arkwright, 25 June 1882, HARC, A63/IV/42/29.
199. J. R. Bailey to J. H. Arkwright, n.d. 1884, HARC, A63/IV/42/33.
200. Ibid.
201. "Concealed Connections," *Livestock Journal* 21 (10 April 1885): 351.

Chapter Four

1. Holford, *New Zealand's Own Breed*, 3.
2. Ibid., 11.
3. Holford, *New Zealand's Contribution*, 11.
4. Holford, *New Zealand's Own Breed*, 12.
5. Ibid., 10.
6. Ibid., 11.
7. See chapter 2.
8. Holford, *New Zealand's Contribution*, 10.
9. Ibid., 10.
10. "Mutton Cutlets," *New Zealand Farmer* 12, no. 2 (February 1892): 38.
11. Pearce, *County of Berkshire*, 46; see also chapter 2.
12. Holford, *New Zealand's Contribution*, 10.
13. Brooking and Pawson, *Seeds of Empire*.
14. "The Farm: Month of October," *New Zealand Farmer* 11, no. 10 (October 1891): 397.
15. "The Farm: March Month," *New Zealand Farmer* 12, no. 3 (March 1892): 113.
16. Holford, *New Zealand's Own Breed*, 11.
17. Rebecca Woods, "Breed, Culture, and Economy"; Rebecca Woods, "From Colonial Animal to Imperial Edible."
18. David Jones, "New Zealand Trade," 119.
19. Franklin, *Dolly Mixtures*, 122.

20. Ibid., 120.

21. Elinor Melville describes this as an "ungulate irruption." Melville, *Plague of Sheep*, esp. 6–9. See also Flannery, *Future Eaters*. For the ecological transformation of colonies, see Virginia Anderson, *Creatures of Empire*; Cronon, *Changes in the Land*; Crosby, *Columbian Exchange*; and Crosby, *Ecological Imperialism*.

22. Muir, *Broken Promise*.

23. Formal colonization in Australia dates to 1788; in New Zealand, the Treaty of Waitangi (1840) marks the still-contested onset of British sovereignty.

24. "Wool," *New Zealand Country Journal* 2, no. 3 (May 1878): 185.

25. Juliet Clutton-Brock, *Walking Larder*.

26. Stringleman and Peden, "Sheep Farming," http://www.TeAra.govt.nz/en/sheep-farming/page-2.

27. Mitchell, *Abstract of Statistics*, 192.

28. Quoted in "Wool," 187.

29. From its paltry initial offering, the combined export from the Australian colonies' annual clip rose to nearly 100,000 in 1820, to more than three-quarters of a million in 1840, to almost six million in 1850, to two and a quarter million twenty-five years later. Armstrong and Campbell, *Australian Sheep Husbandry*, 61. See also Mitchell, *Abstract of Statistics*, 193.

30. Armstrong and Campbell, *Australian Sheep Husbandry*, 1.

31. Merinos from the state of Vermont, in particular, were popular enough in Australia to be considered a craze in the 1880s. Graham, *Australian Merino*, 13, 20; "Vermont Merinos in Australia," *New Zealand Farmer* 11, no. 12 (December 1891): 492; Rebecca Woods, "Green Mountain Merinos."

32. Graham, *Australian Merino*, 9.

33. "The Australian Meat-Trade," *Chambers's Journal*, 21 April 1894, 246.

34. Robin, *How a Continent Created a Nation*; Melville, *Plague of Sheep*.

35. Holford, *Contribution to the Sheep World*, 10.

36. "District Reports: Wellington Province, Wanganui," *New Zealand Farmer* 12, no. 11 (November 1892): 457.

37. Pawson and Brooking, "Introduction," 3. See also Beattie, O'Gorman, and Henry, *Climate, Science, and Colonization*.

38. "Discovery of Lost Sheep," *New Zealand Country Journal* 4, no. 4 (July 1880): 226.

39. "Frozen Food," *Chambers's Journal*, 14 July 1883, 437.

40. Armstrong and Campbell, *Australian Sheep Husbandry*, 8.

41. "The Flocks of the Empire," *New Zealand Farmer* 21, no. 2 (February 1901): 46.

42. Ibid.

43. Ibid.; *Statistical Abstract 1889 to 1903*, 2.

44. J. R., "Transportation of Live Stock, Part II: Public Health and Public Morals," *Livestock Journal and Fancier's Gazette* 2 (6 August 1875): 353.

45. The population of human colonists is given by *Statistical Abstract 1876 to 1890*, 5. The numbers of sheep in the Australian colonies and New Zealand are compiled from Grant, "Australian Meat Industry," 1:33, 35; and Evans, *Agricultural and Pastoral Statistics*, 31, 7.

46. In 1881, the population of settlers, or *pakeha*, stood at approximately 440,000. The Maori population was roughly 45,000 in the same year; it "reached its nadir of 42,000" in 1892. Brooking and Pawson, "Contours of Transformation," 13. See also McLintock, *Encyclopedia of New Zealand.*

47. David Jones, "New Zealand Trade," 130.

48. *Cyclopedia of New Zealand*, 250–51.

49. "Export of Frozen Meat," *Timaru Herald*, 24 March 1881, 8.

50. The conversion of numbers of sheep to pounds of meat is based on Holmes's calculations, which estimated twenty sheep per ton, or 100,000 tons of meat for two million sheep. "Frozen Meat Export Company," *North Otago Times*, 28 February 1881, 2.

51. Perren, *Meat Trade in Britain*, 3.

52. Davidson, *Establishment of the Frozen-Meat Trade*, 10. National Archives of Scotland GD435/614/6. Similar measures were reported resorted to in Argentina. Waters, *Clipper Ship*, 53.

53. "Australian Mutton," *All the Year Round*, 12 September 1868, 319.

54. Some in New Zealand, in fact, looked forward to the day when Australia's "large city populations capable of consuming enormous quantities of such commodities as New Zealand is particularly fitted to produce" would become "the best customers our cultivators of the soil will have." "New Zealand and Intercolonial Federation," *New Zealand Farmer* 11, no. 4 (April 1891): 145.

55. Collins, "Rural and Agricultural Change," 115; Waters, *Clipper Ship*, 52; Leonard W. Lillingston, "Frozen Food," *Good Words*, January 1898, 238.

56. "Australian Mutton," *All the Year Round*, 1868, 319–20.

57. Collins, "Food Supplies and Food Policy," 37.

58. "Scientific Notes," *The Graphic*, 3 December 1881, repr. in *Haslam's Patent Dry Air Refrigerators*, 10. Derbyshire Records Office (hereafter, DRO), D1333 Z/Z 5.

59. "The Australian Meat-Trade," *Chambers's Journal*, 21 April 1894, 246; Perren, *Meat Trade in Britain*, 70–74.

60. Davidson, *Establishment of the Frozen-Meat Trade*, 33; Waters, *Clipper Ship*, 52.

61. Ramsay, "World's Refrigerated Meat Traffic," 1721.

62. "Sir Alfred Haslam, KT., J.P.: A Sketch of His Career," in *The Queen's State Visit to Derby May 21st, 1891*, 140. DRO, D1333 Z/Z 8.

63. Perren, *Meat Trade in Britain*, 3. Domestic net production was growing at a modest rate of 1.5 percent per annum over the second half of the nineteenth century. Collins, "Rural and Agricultural Change," 116.

64. Higgins, "Mutton Dressed as Lamb," 175–76.

65. Collins, "Rural and Agricultural Change," 110.

66. Ibid., 98, 109.

67. Quoted in "The Diminution of Live Stock," *New Zealand Country Journal* 2, no. 3 (May 1878): 170.

68. "Our Meat-Supply," *Chambers's Journal*, 28 November 1868, 759.

69. "Our Meat-Supply," *Chambers's Journal*, 26 August 1899, 615–16; Perren, *Meat Trade in Britain*, 3. Prior to World War I, meat consumption in Britain peaked in the first five years of the twentieth century at 132 lbs. per capita.

70. Collins, "Food Supplies and Food Policy," 35; Higgins, "Mutton Dressed as Lamb," 166. See also chapter 5.

71. Craigie, "Twenty Years' Change in Our Foreign Meat Supplies," *Journal of the Royal Agricultural Society of England* 23, 2nd ser. (1887): 472.

72. "American Meat," *Saturday Review of Politics, Literature, Science and Art* 52, no. 1366 (1881): 811.

73. "Imported Beef and Mutton," *Chambers's Journal*, 22 April 1871, 253.

74. "Our Meat-Supply," *Chambers's Journal*, 1868, 759; Ritvo, "Mad Cow Mysteries," 99–100; Ritvo, *Platypus and Mermaid*, 194–97. Otter, "Civilizing Slaughter," 89; Rogers, *Beef and Liberty*.

75. "Flocks of the Empire," 46.

76. Ibid.

77. "Scientific Notes," *Haslam's Dry Air Refrigerators*, 10.

78. Gordon H. Campbell, quoted in *Proceedings of Fourth International Congress of Refrigeration*, 32.

79. Davidson, *William Soltau Davidson*, 10; "Australian Mutton," *All the Year Round* 20, no. 490, (12 September 1868): 319.

80. "Australian Meat-Trade," *Chambers's Journal*, 1894, 246.

81. Starting in the late 1860s, Thomas Mort and James Harrison in Australia began experimenting with freezing meat, but were unable to successfully ship it in its frozen state. Waters, *Clipper Ship*, 52.

82. Wallis-Tayler, *Refrigeration*, 21.

83. Ibid., 366; Waters, *Clipper Ship*, 53; Ramsay, "World's Refrigerated Meat Traffic," 1722; Cronon, *Nature's Metropolis*, 230–47; "American Meat; How It Comes," reprinted from *The Farmer* in the *New Zealand Country Journal* 1, no. 3 (July 1877): 194–95.

84. These relied on the dry air process of refrigeration, in which the compression of atmospheric air is used to cool an insulated chamber. The alternative, chemical refrigeration, relied on substances like anhydrous ammonia. Chemical refrigeration allowed for a more efficient heat cycle, but the chemicals used for it were flammable, and did "injurious action" upon the copper pipes that were needed to desalinate sea water for "marine refrigeration." Wallis-Tayler, *Refrigeration*, 48, 211, 396.

85. "Australian Meat-Trade," *Chambers's Journal* (1894): 246.

86. Ramsay, "World's Frozen Meat Trade," 4. In 1873, an attempt was made to ship frozen meat from Melbourne to London, but it "turned out a failure." Wallis-Tayler, *Refrigeration*, 2.

87. "Exportation of Butter," *Star*, 17 February 1881, 3.

88. Thomas Mackenzie, quoted in *Proceedings of Fourth International Congress of Refrigeration*, 1:46.

89. Critchell and Raymond, *History of the Frozen Meat Trade*, 39.

90. "Notes and Comments," *Otago Daily Times*, 17 December 1881, 7.

91. "Christmas Relish," *Otago Daily Times*, 26 December 1881, 3.

92. Only one carcass was condemned. Critchell and Raymond, *History of the Frozen Meat Trade*, 42; "Our Meat-Supply," *Chambers's Journal*, 1899, 616.

93. Davidson, *William Soltau Davidson*, 37; Critchell and Raymond, *History of the Frozen Meat Trade*, 415; Waters, *Clipper Ship*, 51–57.

94. Wallis-Tayler, *Refrigeration*, 7.

95. By 1910, the United Kingdom had the capacity to store more than eight million sheep carcasses. Critchell and Raymond, *History of the Frozen Meat Trade*, 418–19. The East and West India Dock Company and the London and St. Katharine Dock Company led the establishment of "public refrigerated accommodation." Broodbank, "Development of Refrigerated Accommodation," 1705.

96. Wallis-Tayler, *Refrigeration*, 6.

97. "Arrival of the Sailing Ship Mataura," *European Mail*, 5 October 1882, repr. in *Haslam's Dry Air Refrigerators*, 12.

98. Davidson, *Establishment of the Frozen-Meat Trade*, 15–16.

99. "Frozen Meat Export Company," 2.

100. This risk of exposure was also a problem at the other end of the journey, where transfer "from the vessel to the cold stores on land, and subsequent distribution by road or rail to the retailers," offered ample opportunity for exposure to higher temperatures. Wallis-Tayler, *Refrigeration*, 6, 365.

101. Alexander Bruce, "The New Zealand Frozen Meat Trade," *Australasian Pastoralists' Review* 2, no. 11 (14 January 1893): 1024.

102. "Export of Frozen Meat," *Otago Daily Times*, 24 March 1881, 7.

103. Bruce, "New Zealand Frozen Meat Trade," 1024.

104. "The Frozen Meat Industry," *Australasian Pastoralists' Review* 3, no. 2 (15 April 1893): 72.

105. "A Visit to the Australian Frozen Meat Company's Works," *Leisure Hour*, September 1882, 561.

106. Bruce, "New Zealand Frozen Meat Trade," 1025.

107. "Frozen Meat Export Company," 2.

108. For race and the colonization of Australia, see Warwick Anderson, *Cultivation of Whiteness*.

109. "Australian Meat-Trade," *Chambers's Journal*, 1894, 247.

110. Lillingston, "Frozen Food," 241.

111. Ibid., 241; Wallis-Tayler, *Refrigeration*, 285.

112. Wallis-Tayler, *Refrigeration*, 374, 379.

113. "Australian Refrigerated Meat," *Daily News*, 5 October 1881, repr. in *Haslam's Dry Air Refrigerators*, 6.

114. "Arrival of Frozen Meat from Australia," *Daily News*, 24 October 1881, repr. in *Haslam's Dry Air Refrigerators*, 7.

115. Lillingston, "Frozen Food," 238.

116. "Our Meat-Supply," *Chambers's Journal*, 1899, 616.

117. "Visit to the Australian Frozen Meat Company," 560.

118. Craigie, "Twenty Years' Change," 465.

119. "American Meat," *Saturday Review of Politics, Literature, Science and Art* 52, no. 1366 (1881): 812.

120. Lillingston, "Frozen Food," 237.

121. Ernest E. Williams, "The Foreigner in the Farmyard," *New Review* 16, no. 93 (February 1897): 149.

122. Ibid.

123. "Annual Statement of the Trade of the United Kingdom with Foreign Countries and British Possessions for the Year 1885," *Quarterly Review* 165, no. 329 (July 1887): 54–55; Williams, "Foreigner in Farmyard," 150.

124. Williams, "Foreigner in Farmyard," 150; Lillingston, "Frozen Food," 238; Ritvo, *Platypus and Mermaid*, 194.

125. Lillingston, "Frozen Food," 238.

126. "The Australian Meat-Trade," *Chambers's Journal* (1894): 247.

127. Lillingston, "Frozen Food," 238.

128. Higgins, "Mutton Dressed as Lamb," 173, 175, 177.

129. House of Lords, *Report on Marking Foreign Meat*.

130. Higgins, "Mutton Dressed as Lamb," 167–71.

131. House of Lords, *Report on Marking Foreign Meat*, xi. Economic historian David M. Higgins has conducted a detailed analysis of the evidence given to the Select Committee, and concluded that not only was fraud less prevalent than contemporaries supposed, its effects were also less pernicious. Had fraud existed at a significant scale, Higgins argues, the price differential between meat of foreign origin, including colonial, and domestically produced meat would have narrowed over time. That this did not occur suggests a relatively low degree of fraud in the marketplace. What misrepresentation existed, Higgins concludes, was practiced over a relatively short span of time in the early years of the trade. Moreover, outrage over the misrepresentation of the point of origin of meat expressed an objection to the act of fraud itself, not necessarily a prejudice against foreign or colonial meat. That is, consumers objected to being sold a false article (colonial meat passed off as British), not necessarily to colonial meat per se. Britons wished "to exercise their patriotic preference in favour of domestic meat," and misrepresentation of colonial mutton as British prevented them from doing so. Higgins, "Mutton Dressed as Lamb" 182, 176, 174.

132. As Higgins argues in "Mutton Dressed as Lamb."

133. "Fraudulent Dealings with New Zealand Mutton," *New Zealand Farmer* 11, no. 9 (September 1891): 357.

134. Ibid.

135. "The Frozen Meat Industry," *Australasian Pastoralists' Review* 3, no. 2 (15 April 1893): 73.

136. "Fraudulent Dealings," 357.

137. *Australasian Pastoralists' Review* 2, no. 10 (15 December 1892): 953.

138. "Fraudulent Dealings," 357.

139. "Pure-Bred Hampshire Downs," *New Zealand Farmer* 12, no. 1 (January 1892): 3.

140. "Frozen Meat Industry," *Australasian Pastoralists' Review*, 72.

141. "Export Only Good Mutton," *New Zealand Farmer* 11, no. 9 (September 1891): 371.

142. "River Plate and New Zealand Mutton," *New Zealand Farmer* 12, no. 1 (January 1892): 4.

143. "Export Only Good Mutton," 371.

144. *Australasian Pastoralists' Review* 2, no. 11, (14 January 1893): 993. Australians worried that, while "Australian meat sold as Australian finds ready market both in London and the country towns, but every now and then a shipment of inferior mutton comes in from other places, the meat is sold as Australian, the public are dissatisfied, and will not buy again for some time." *Australasian Pastoralists' Review*, 15 March 1893, 37.

145. Lillingston, "Frozen Food," 239.

146. "How Down Mutton Went Down," *New Zealand Farmer* 12, no. 9 (September 1892): 370.

147. *Australasian Pastoralists' Review* 3, no. 2 (15 April 1893): 72.

148. William Darley, "Mutton Cutlets," *New Zealand Farmer* 12, no. 2 (February 1892): 38.

149. "Down and Lincoln Breeds of Sheep," *New Zealand Farmer* 12, no. 6 (June 1892): 249.

150. "The Frozen Meat Trade in Victoria and New Zealand," *New Zealand Farmer* 20, no. 5 (May 1900): 193.

151. Taylor White, "Cross-Breeding of Sheep," *New Zealand Farmer* 12, no. 5 (May 1892): 198.

152. Corin, "The Management of Sheep on Small Farms," *New Zealand Farmer* 12, no. 5 (May 1892): 197.

153. Worgan, *County of Cornwall*, 149.

154. Taylor White, "On Cross Breeding Sheep," *New Zealand Farmer* 12, no. 4 (April 1892): 157.

155. Corin, "Management of Sheep," 197.

156. T. H. Anson, "On Sheep," *New Zealand Country Journal* 1, no. 3 (July 1877): 190.

157. "The Farm: September Month," *New Zealand Farmer* 11, no. 9 (September 1891): 1.

158. "The Farm: November Month," *New Zealand Farmer* 11, no. 11 (November 1891): 445.

159. Butler, *A First Year in Canterbury Settlement*, 36.

160. Anson, "On Sheep," 190.

161. John McBeath, "Cross-Breeding of Sheep," *New Zealand Country Journal* 1, no. 4 (October 1877): 267. Italics in original.

162. "Breeding for Wool in the North," *New Zealand Farmer* 12, no. 1 (January 1892): 4.

163. "Farm: Month of October," 397.

164. White, "On Cross Breeding Sheep," 157.

165. "Rusticus," "Stud Breeding," *Australasian Pastoralists' Review* 3, no. 2 (15 April 1893): 76.

166. Ibid.

167. Corin, "Management of Sheep," 197.

168. W. Weddel, quoted in "The Mutton of Most Value in London Markets," *New Zealand Farmer* 12, no. 12 (December 1892): 476.

169. "Visit to the Australian Frozen Meat Company," 560.

170. Lillingston, "Frozen Food," 239.

171. See also Ritvo, "Possessing Mother Nature."

172. Holford, *New Zealand's Own Breed*, 11.

173. John Roberts, "Crossbred Sheep in New Zealand," *Australasian Pastoralists' Review* 2, no. 9, (15 November 1892): 220. See also "The Intercolonial Stock Conference," *Australasian Pastoralists' Review* 2, no. 9 (15 November 1892): 903–4.

174. Roberts, "Crossbred Sheep," 220.

175. White, "On Cross Breeding Sheep," 156.

176. "Cross-Bred Sheep," *New Zealand Country Journal* 1, no. 4 (October 1877): 269.

177. Ibid.

178. White, "Cross-Breeding of Sheep," 198.

179. "Probable Changes in New Zealand Sheep-Breeding," *New Zealand Farmer* 12, no. 11 (November 1892): supp., 4.

180. "English and New Zealand Bred Lincolns," *New Zealand Farmer* 20, no. 1 (January 1900): 18.

181. "Probable Changes," 4; White, "On Cross Breeding Sheep," 156.

182. "Southdown Prize Ram," *New Zealand Farmer* 11, no. 10 (October 1891): 400.

183. "The Dry Air Refrigerator, or Freezing Machine," *British Mail*, April 1882, reprinted in *Haslam's Dry Air Refrigerators*, 11.

184. "Downs V. Lincolns for Crossing," *New Zealand Farmer* 12, no. 7 (July 1892): 278. Also reprinted in "Sheep Breeding," *Australasian Pastoralists' Review* 2, no. 8 (15 October 1892): 881.

185. Ibid.

186. White, "On Cross Breeding Sheep," 156.

187. Ibid.

188. W. Weddel, quoted in "Mutton of Most Value," 476.

189. Anson, "On Sheep," 190.

190. "District Reports: Canterbury Province, North and Mid," *New Zealand Farmer* 11, no. 11 (November 1891): 468.

191. "Show Reports: Canterbury Agricultural and Pastoral Association's Metropolitan Show," *New Zealand Farmer* 11, no. 12 (December 1891): 522.

192. "District Reports: Canterbury Province," 468.

193. Little established his pure-breeding cross from the Romney breed and merino sheep "before anyone else had thought of such a thing." Little, *Story of the Corriedale*, 3.

194. Davidson, *William Soltau Davidson*, 23.

195. Davidson, *Establishment of the Frozen-Meat Trade*, iv.

196. In fact, controversy surrounding early efforts led parties involved to publish memoirs detailing competing claims to primacy. "Corriedale" was eventually settled upon as the name for the fixed cross in honor of Little's initial efforts as the manager of an Otago estate by that name. Davidson and the New Zealand and Australia Land Company advocated for "Southern Cross" as a moniker for the type. Davidson, *William Soltau Davidson*; Little, *Story of the Corriedale*.

197. "District Reports: Wellington Province, Wanganui," *New Zealand Farmer* 12, no. 11 (November 1892): 457.

198. Ibid.
199. White, "On Cross Breeding Sheep," 156.
200. Ibid.
201. "Of the improved Leicester," White claimed, "we have no certain knowledge whether Bakewell used a cross or no. They are said to be descended from the old Teeswater breed." White, "On Cross Breeding Sheep," 157.
202. "Selecting Sheep for a Breeding Flock," *New Zealand Farmer* 11, no. 4, (April 1891): 133.
203. White, "On Cross Breeding Sheep," 157.
204. "Selecting Sheep," 133.
205. *Australasian Pastoralists' Review* 3, no. 1 (15 March 1893): 4.
206. White, "On Cross Breeding Sheep," 157.
207. Ibid.
208. Ibid.
209. By and large, interested parties in New Zealand fell out along the lines of Down breeds versus the longwools, in recognition of the changing preferences of Britons, and the need to cater to metropolitan taste if New Zealand was to maintain primacy in the frozen meat trade. See, for example, "Pure-Bred Hampshire Downs," 3; "Longwool and Down Mutton Sheep," *New Zealand Farmer* 12, no. 1 (January 1892): 21; "Down and Lincoln Breeds of Sheep," *New Zealand Farmer* 12, no. 6 (June 1892): 249; "The Down v. Lincoln Question," *New Zealand Farmer* 12, no. 6 (June 1892): 250.
210. White, "On Cross Breeding Sheep," 157.
211. Ibid.
212. Ibid.
213. Ibid.
214. Ibid.
215. Holford, *New Zealand's Own Breed*, 2.
216. "District Reports: Palmerston North," *New Zealand Farmer* 12, no. 11 (November 1892): 458.
217. Holford, *New Zealand's Own Breed*, 4.
218. Ibid., 3.
219. Ramsay, "World's Frozen Meat Trade," 5.
220. Holford, *Contribution to the Sheep World*, 11.
221. Holford, *New Zealand's Own Breed*, 17.

Chapter Five

1. "Hereford Cattle," *Maitland Mercury & Hunter River General Advertiser*, 3 October 1885, 21.
2. "Hereford Cattle," *Livestock Journal and Fancier's Gazette* 2 (12 November 1875): 689.
3. Rogers, *Beef and Liberty*; Shapin, "You Are What You Eat."
4. "Hereford Cattle," *Maitland Mercury*, 1885, 21.
5. T. S., "On the Choice and Management of Dairy Stock, with a Few Observations on the Best Methods of Rearing Calves," *Agricultural Magazine, or, Farmers' Monthly Journal* 3 (July 1808): 7.

6. Weston, "General Remarks on the Shew of Fat Cattle in Smithfield," *Commercial and Agricultural Magazine* 5, no. 29 (December 1801): 383.

7. T. S., "Choice and Management of Dairy Stock," 7.

8. "British Breeds of Cattle," *Livestock Journal* 21 (22 May 1885): 495.

9. Ibid.

10. "Hereford Cattle," *Livestock Journal*, 1875, 689.

11. "Hereford Cattle," *Maitland Mercury*, 1885, 21.

12. "Hereford Cattle," *Brisbane Courier*, 13 December 1882, 3.

13. "Hereford Cattle," *Maitland Mercury*, 1885, 21.

14. "Hereford Cattle," *Livestock Journal*, 1875, 689.

15. Duckham's essay "A Lecture on the History, Progress, and Comparative Merits of the Hereford Breed of Cattle" was first delivered at the Royal Agricultural College at Cirencester, 4 December 1863. In 1869, he had it reprinted in volume 6 of *Eyton's Herd Book of Hereford Cattle*. The bulk of it made its way into the Tasmanian weekly, *The Mercury*, in 1872, by way of *The Field*, a major British publication concerned with sporting and agricultural pursuits. The full text of the lecture was also included in the first volume of the *New Zealand Herd Book* in 1886. See Duckham, "History, Progress, and Comparative Merits"; "The Hereford Breed of Cattle," *Mercury*, 3 May 1872, 3; *New Zealand Herd Book*.

16. "Hereford Cattle," *Livestock Journal*, 1875, 689.

17. "Herefords in Westmeath," *Livestock Journal* 2 (3 September 1875): 450.

18. Ibid. In an interesting conflation of the practitioner and his subject—not unlike that observed by Rebecca Cassidy among contemporary thoroughbred horse breeders—commentators often attributed a breeder's skill to heredity. The son of another prominent Hereford breeder, William Tudge, was said to have "inherited his father's taste for fine cattle," and in the Ashburner family, the production of several good breeders of Shorthorns occasioned the *Livestock Journal* to remark that, in that family, "the taste for Shorthorns is thus hereditary." Cosmo, "Among the Herefords: Mr. Tudge's Herd at Leinthall," *Livestock Journal* 21 (1 May 1885): 424; "Shorthorns for California," *Livestock Journal* 2 (29 October 1875): 642; Cassidy, *Sport of Kings*.

19. R. W. Reynall, quoted in Duckham, "History, Progress, and Comparative Merits," 27.

20. Reynall, quoted ibid.

21. Samuel Gilliland, quoted ibid., 26.

22. "Hereford Cattle," *Maitland Mercury*, 1885, 21.

23. Ibid.

24. Mr. Lumsden, quoted in Duckham, "History, Progress, and Comparative Merits," 26.

25. Ibid., 22.

26. John Murrison, quoted ibid, 25–26.

27. James Mappower, quoted ibid, 23.

28. Youatt, *Cattle*, 32.

29. "Hereford Cattle," *Maitland Mercury*, 1885, 21.

30. Ibid.

31. Duckham, "Breeding and Management," 3. Paper originally given at the Breconshire Chamber of Agriculture, 2 January 1869.

32. J. R., "Foreign and Irish Live Stock and Disease," *Livestock Journal* 2 (11 June 1875): 187.

33. "Imported Cattle and Disease," *Livestock Journal* 2 (27 August 1875): 424.

34. Ibid.

35. J. R., "Foreign and Irish Live Stock and Disease," 187.

36. Collins, "Rural and Agricultural Change," 111; Collins, "Food Supplies and Food Policy," 35. Ireland occupied an uneasy position in the trade and was only inconsistently considered a "foreign" source of meat.

37. Ibid.

38. Perren, *Meat Trade in Britain*, 153. Consumption of foreign sheep meat stood at 45 percent in the same year.

39. Ibid., 114.

40. Ibid., 131. The United States surpassed European imports with 204,467 to 182,572 live cattle in 1880, fell as low as 80,023 to 261,055 in 1882, rose again to surpass European imports in 1885 with 206,350 to 164,936, and remained ahead of Europe for the rest of the century. Ibid., 131, 164.

41. "Hereford Cattle," *South Australian Register*, 14 December 1877, 3. Environmental historians have examined the significance of the development of this industry for American industry, and its ecological consequences. See, in particular, Cronon, *Nature's Metropolis*; White, *Railroaded*.

42. Grundy, "Hereford Bull," 72. The *Quarterly Review* noted in 1887 that during the North American ranching boom of the early 1880s, many "British 'tenderfeet' were induced to invest a great deal of capital in the business," and according to Don Worcester, the Prairie Cattle Company of West Texas was "the mother of the British companies," and "partly responsible for triggering Britons' hasty and incautious investment in ranching ventures in the late 1870s and 1880s." "Our Meat Supply," *Quarterly Review*, July 1887, 49; Worcester, *Texas Longhorn*, 57.

43. "English Stock in Kansas," *Livestock Journal* 2 (19 November 1875): 713.

44. Ibid.

45. Steam transport first brought the prairies into the productive orbit of American metropoles—New York, Chicago, and others. Their extension to London was both coeval with and dependent on this development. Belich examines the global dimensions of these developments in detail. See Cronon, *Nature's Metropolis*; White, *Railroaded*; Belich, *Replenishing the Earth*.

46. "English Stock in Kansas," 688.

47. W. H. Sotham, "Colonial Agriculture," *Farmer's Magazine* 55 (January 1879): 21.

48. To contemporary American observers, transforming prairies into pasture promised profit, but the cost of this endeavor, as any number of environmental historians have shown, was irrevocable ecological and social change. See especially Isenberg, *Destruction of the Bison*, and Jacoby, *Crimes against Nature*, for the more immediate ecological consequences in the nineteenth century. For some of the long-term ecological consequences of converting the prairies to farm and

pasture land, see Worster, *Dust Bowl*. Terry G. Jordan sees in the transition to cattle ranching, a romanticized way of life being instituted at the expense of Native Americans. Jordan, *North American Cattle-Ranching Frontiers*, 7.

49. Cronon's analysis of the processes and consequences of the commercialization of agriculture in the American West and Midwest is the most comprehensive; Jordan's analysis of cattle ranching the most detailed by region. See Cronon, *Nature's Metropolis*; Jordan, *North American Cattle-Ranching Frontiers*.

50. The favorable rates on North Atlantic freight, as well as the ease of finding cargo for the return trip to Montreal or New York, had a material influence on the development of the meat trade between North America and Great Britain. See Harley, "Steers Afloat."

51. Perren, *Meat Trade in Britain*, 114.

52. Ibid.

53. "London Dead Meat Market," *Livestock Journal* 21 (6 February 1885): 130.

54. Ibid.

55. Ibid.

56. By 1890, U.S. chilled imports stood at 1.7 million hundredweights, while live imports had risen to 384,639. Perren, *Meat Trade in Britain*, 116, 170, 164.

57. "Wyoming Cattle," *Livestock Journal* 21 (13 February 1885): 147.

58. Peter King, "Genetic Diversity of Traditional British Breeds of Beef Cattle," *Ark* 21, no. 1 (Spring 1996): 27.

59. Ramsay, "World's Frozen and Chilled Meat Trade," 5.

60. "Wyoming Cattle," 147.

61. Ibid.

62. "Imported Beef and Mutton," *Chambers's Journal*, 21 April 1877, 254.

63. Perren, *Meat Trade in Britain*, 115.

64. Not even slaughtering and chilling prior to shipment entirely eliminated the risks of the voyage. Contemporary observation suggested that "the meat from animals slaughtered on their arrival in Liverpool is better than the dead meat imported from America, because the dead meat has suffered inevitable injury from being knocked about during its transport across the sea." "American Meat," *Saturday Review*, 13 December 1881, 812.

65. Derry, *Bred for Perfection*, 34.

66. Grundy, "Hereford Bull," 87.

67. Joan Grundy remarks that ranges were "stocked... at a phenomenal rate" in the 1880s. Ibid., 73.

68. MacDonald, *Food from the Far West*, 268.

69. Grundy, "Hereford Bull," 73.

70. "The American Cattle and Dead Meat Industry," *Livestock Journal* 21 (30 January 1885): 102. For the history of livestock animals in colonial America, see Virginia Anderson, *Creatures of Empire*.

71. Sotham, "Colonial Agriculture," 21.

72. "American Cattle and Dead Meat," 102; Worcester, *Texas Longhorn*; Jordan, *North American Cattle-Ranching Frontiers*.

73. MacDonald, *Food from the Far West*, 268. Even Worcester concedes that the "old-time Longhorns" were "not the most handsome of bovines." Worcester, *Texas Longhorn*, 4.

74. MacDonald, *Food from the Far West*, 268.

75. Sotham, "Colonial Agriculture," 21.

76. "Wyoming Cattle," 147.

77. MacDonald, *Food from Far West*, 268–69.

78. "Wyoming Cattle," 147.

79. Ibid.

80. "Imported Beef and Mutton," 254.

81. Perren, *Meat Trade in Britain*, 160.

82. "Wyoming Cattle," 147.

83. "The American Cattle and Dead Meat Industry," *Livestock Journal*, 102.

84. "Stock for Nova Scotia," quoted from the *St. John Daily Telegraph* in *Livestock Journal* 2 (19 November 1875): 718.

85. Worcester, *Texas Longhorn*, 65. Shorthorns and Herefords were introduced to the United States in the 1820s, Aberdeen-Angus cattle in the 1860s. Shorthorns "made the earliest headway with rapid expansion in the eastern states between 1866 and 1878." Grundy, "Hereford Bull," 74.

86. Derry, *Ontario's Cattle Kingdom*.

87. The entire number of cattle sold at auction in the United States in the year 1884 was approximately 7,500, 4,383 of which were Shorthorns and only 314 of which were Herefords (although that was up from 112 the year before). "Herd Intelligence," *Livestock Journal* 21 (23 January 1885): 81.

88. "Cattle of the Various Breeds as Beef Producers," *Farmer's Magazine* 55 (February 1879): 99.

89. Ibid.

90. Grundy, "Hereford Bull," 74. Worcester recounts the story of an early effort to import "blooded stock" to Texas, in which a cattleman had two cows and a bull of the Shorthorn breed hauled in wagons from the port of New Orleans. Another rancher was said to have remarked on the occasion that "a man has no business with cows that can't light out and walk from New Orleans." Worcester, *Texas Longhorn*, 62.

91. Jordan, *North American Cattle-Ranching Frontiers*, 7–10.

92. "Remarks and Observations on Different Sorts of Cattle," *Agricultural Magazine* 7 (November 1810): 325.

93. Sotham, "Colonial Agriculture," 21.

94. Ibid.

95. "Hereford Cattle," *Brisbane Courier*, 1882, 3.

96. Belich, *Replenishing the Earth*.

97. "Hereford Cattle," *Livestock Journal* (1875): 689.

98. J. Edwards, quoted in Duckham, "History, Progress, and Comparative Merits," 29.

99. F. W. Stone, quoted ibid., 28.

100. "Among the Herefords: Current Notes," *Livestock Journal* 21 (8 May 1885): 446; "Hereford Cattle," *Livestock Journal* (1875): 689.

101. Grundy, "Hereford Bull," 76.

102. Derry, *Bred for Perfection*, 34; Jordan, *North American Cattle-Ranching Frontiers*, 201; Grundy, "Hereford Bull," 72.

103. "Hereford Cattle," *South Australian Register*, 1877, 3; "Hereford Cattle in America," *Maitland Mercury*, 26 May 1883, 6.

104. A. B., "The Hereford Cattle Trade in America," *Livestock Journal* 21 (30 January 1885): 101.

105. "Cattle as Beef Producers," 99.

106. Jordan, *North American Cattle-Ranching Frontiers*, 274.

107. Ibid., 9.

108. Grundy, "Hereford Bull," 71.

109. Worcester, *Texas Longhorn*, 64.

110. Ritvo, "Possessing Mother Nature," 417.

111. Even a quarter-bred Hereford bull color-marked its offspring. Grundy, "Hereford Bull," 70, 75.

112. "Hereford Cattle," *South Australian Register*, 1877, 3.

113. "Hereford Cattle," *Capricornian*, 2 January 1892, 11.

114. Ibid.

115. Note that this perception ran counter to the majority of nineteenth-century opinion about the Shorthorn's temperament, which was held elsewhere to be unusually placid.

116. "Hereford Cattle," *Capricornian*, 1892, 11; Beardmore, quoted in "Hereford Cattle," *Brisbane Courier*, 1882, 3.

117. Edwyn Arkwright to John Hungerford Arkwright, 20 July 1885, Herefordshire Archive and Records Centre (hereafter, HARC), A63/IV/21/3.

118. E. Arkwright to J. H. Arkwright, 18 February 1884. HARC, A63/IV/21/3.

119. Ibid., 20 July 1885, HRAC, A63/IV/21/3.

120. Ibid., 13 March 1889, HARC, A63/IV/21/3.

121. Ibid., 13 March 1889; and 18 February 1884.

122. "Various Notes," *Farmer's Magazine* 57 (February 1881): 116.

123. Worcester, *Texas Longhorn*, 67.

124. "Cattle as Beef Producers," 99.

125. Pearce, *County of Berkshire*, 46.

126. "Hereford Cattle," *Brisbane Courier*, 1882, 3.

127. Ibid.

128. Ibid.

129. A. B., "Hereford Cattle Trade in America," *Livestock Journal* (1885): 105.

130. Grundy, "Hereford Bull," 76. By T. L. Miller's count in the year 1883, already by the month of July, more than 1,000 Herefords had been "bot [sic] for the American trade." T. L. Miller to J. H. Arkwright, 19 July 1883, HARC A63/IV/42/31.

131. Four extant ledgers covering the period 1890–1953 are still available at the head office of the Hereford Cattle Society in the city of Hereford.

132. Export Ledger 1890–1901, Hereford Cattle Society.

133. "Hereford Cattle in America," *Maitland Mercury*, 1883, 6.

134. "The Hereford Cattle Outlook," *Launceston Examiner* (Tasmania), 6 April 1881, 2.

135. The same was true of the "grading up" system in Canada. See Derry, *Ontario's Cattle Kingdom*.

136. Worcester, *Texas Longhorn*, 68.

137. Ibid., 57, 68. Goodnight's operation was one of the biggest in the American West, and even his acquisition of a relatively paltry forty "thoroughbred... imported" Hereford bulls was news in Great Britain. "Herd Intelligence: Herefords," *Livestock Journal* 21 (2 April 1885): 330.

138. "Hereford Cattle," *Maitland Mercury*, 1885, 21.

139. Ibid.

140. This represents the lion's share of the 495 total cattle exported to South America in this period, the largest single proportion after the 283 unspecified South American destinations. Hereford Cattle Society, Export Ledger 1890-1901; Grundy, "Hereford Bull," 80-81; "Hereford Cattle in America," *Maitland Mercury*, 1883, 6; "Hereford Cattle," *Maitland Mercury*, 1885, 21.

141. Grundy, "Hereford Bull," 76, 86.

142. Joseph Russell Bailey to John Hungerford Arkwright, 20 May 1884, HARC, A63/IV/42/33.

143. Sotham, "Colonial Agriculture," 21.

144. Ibid.

145. Ibid.

146. "Dispersion of Mr. Knight's Herd at Leinthall," *Livestock Journal* 21 (1 May 1885): 425.

147. Cosmo, "Among the Herefords: The Field Herd," *Livestock Journal* 21 (15 May 1885): 471.

148. Ibid.

149. Ibid.

150. Ibid.

151. Unnamed correspondent, quoted ibid.

152. "British Breeds of Cattle," 495.

153. Hereford Herd Book Society, *Herd Book of Hereford*, 10:viii.

154. Eyton, *Herd Book of Hereford Cattle*, 1:iii.

155. Hereford Herd Book Society, *Herd Book of Hereford*, 10:vii.

156. Ibid.

157. *American Hereford Record*, 1.

158. Ibid.

159. Ibid. The tendency in the British *Herd Book* to "give only the dam's name and the name and number of her sire, and after carrying these dams back for three or four generations, omit the name of the dam" was criticized in the preface to the first volume of the *American Hereford Record*, despite the fact that this was in keeping with the *Herd Book*'s own rules for entry. At the same time, the editors stipulated that "the lack of further information is no discredit to the pedigree," which seemed to negate the very basis of their own complaint. Ibid.

160. T. L. Miller to J. H. Arkwright, 21 July 1883, HARC, A63/IV/42/31.

161. Hereford Herd Book Society, *Herd Book of Hereford*, 10:viii.

162. T. L. Miller to S. W. Urwick, n.d. 1883, HARC, A63/IV/42/31.

163. Miller to Arkwright, 21 July 1883.
164. *American Hereford Record*, 1.
165. Miller to Urwick, n.d. 1883.
166. Ibid.
167. Ibid.
168. T. L. Miller to J. H. Arkwright, 25 July 1883, HARC, A63/IV/42/31.
169. Bailey to J. H. Arkwright, 1883, HARC, A63/IV/42/31.
170. Ibid.
171. Ibid.
172. Hereford Herd Book Society, *Herd Book of Hereford*, 10:viii.
173. J. R. Bailey to J. H. Arkwright, 26 May 1881, HARC, A61/IV/42/26. The "stain of blood" in the United States was in fact carried to 1/32 in the case of African American heritage.
174. This kind of comparison, of course, has a very specific history and politics in nineteenth-century America, and was a characteristic of American slavery. See Johnson, *Soul by Soul*. More generally, see Ritvo, *Platypus and the Mermaid*.
175. Grundy, "Hereford Bull," 79.
176. Ibid., 81. Grundy notes that the Hereford's color-marking ability was as significant in accounting for its popularity in twentieth-century Britain's artificial insemination industry as it had been on the nineteenth-century New World range. Ibid., 83. For the growing popularity of the Hereford breed in postwar Britain, see also Cincinnatus, "Farming Notes: Breed-preference Changes," *Country Life*, 130 (6 July 1961): 43; Simba, "Farming Notes: Breeding for Beef," *Country Life*, 131 (8 March 1962): 555.
177. Ritvo, "Possessing Mother Nature."

Conclusion

1. "Origin and Distribution of the Sheep," *New Zealand Farmer*, 12, no. 6 (June 1892): 237. Reprinted from the *Agricultural Gazette*.
2. Ibid.
3. Contemporary archeozoology supports this proposition. See Juliet Clutton-Brock, *Natural History of Domesticated Mammals*, 74.
4. "Origins and Distribution," 237.
5. Ibid.
6. "Livestock Committee's Report," *New Zealand Farmer* 11, no. 11 (November 1891): 461.
7. Halifax, "Foreword," 5.
8. P. G. Craigie, "Twenty Years' Change in Our Foreign Meat Supplies." *Journal of the Royal Agricultural Society of England* 23, 2nd ser. (1887): 465.
9. Abigail Woods explores some of the impetus to breed livestock for higher yields in "Breeding Cows, Maximising Milk." See also Sayer, "Animal Machines."
10. "Native Breeds—New Rare Breeds Classification," *Ark* 24, no. 2 (Summer 1996): 55.
11. Ibid.; Peter King, "Genetic Diversity of Traditional British Breeds of Beef Cattle," *Ark* 21, no. 1 (Spring 1996): 27.

12. "Native Breeds," 55.

13. King, "Traditional British Breeds," 27. The first breed to have an established herd book was the Shorthorn, in 1822. The *Herd Book of Hereford Cattle* was established in 1846. Coates, *Short-Horned Herd-Book*; Eyton, *Herd Book of Hereford Cattle*.

14. King, "Traditional British Breeds," 27.

15. Ibid.

16. Low, *On the Domestic Animals of the British Islands*, 116.

17. King, "Traditional British Breeds," 27.

18. Youatt, *Complete Grazier*, 10. See also chapter 1.

19. King, "Traditional British Breeds," 27.

20. Ibid.

21. Ibid.

22. Ibid.

23. In June 1960, John Cumber wrote in *Country Life* magazine that for "possibly a century or more, Britain has been known as the stud farm of the world and British breeds of livestock have been, if not the only ones in the world, certainly the foremost." John Cumber, "Future Trends in Livestock Breeding," *Country Life*, 77 (30 June 1960): 1484.

24. "Native Breeds," 55.

25. Ibid.

26. Peter King erroneously dates this introduction to 1968 (King, "Traditional British Breeds," 27). The question of importing Charolais bulls to Great Britain was a matter of controversy, the risk of disease at least as worrying as the potential for introgression in British beef breeds like the Hereford, Aberdeen-Angus, or Red Poll. Thirty bulls were purchased in October of 1961 at an average cost of £560, and by March 1962 the Milk Marketing Board had made the semen of sixteen available to interested farmers. See Cincinnatus, "Farming Notes: The Charollais Project," *Country Life* 80 (26 October 1961): 1027; Cincinnatus, "Farming Notes: Charollais Challenge," *Country Life* 131 (15 March 1962): 619. Also "Charollais Bulls," *Country Life* 78 (15 September 1960): 538; Cincinnatus, "Farming Notes: The Charollais Controversy," *Country Life* 78 (22 December 1960): 1569; Cincinnatus, "Farming Notes: The Charollais Bull Test," *Country Life* 79 (18 May 1961): 1175; Cincinnatus, "Farming Notes: Charollais Bulls," *Country Life* 80 (27 July 1961): 213; Cincinnatus, "Farming Notes: No Charollais for Scotland?," *Country Life* 80 (9 November 1961): 1159; and Cincinnatus, "Farming Notes: Charollais Bulls," *Country Life* 80 (28 December 1961): 1645.

27. In 1960, *Country Life* magazine reported that there seemed "to be a perfect mania these days to look abroad for foreign breeds of livestock to import." "Farming Notes: Interest in Pietrain Pigs," *Country Life* 77 (9 June 1960): 1341.

28. Edward Hart, "The Traditional Hereford," *Ark* 27, no. 2 (Summer 1999): 64; Peter Symonds, quoted in "Traditional Hereford Breeders Get Together," *Ark* 24, no. 3 (Autumn 1996): 64.

29. The Hereford Cattle Society began distinguishing "traditional" entries to the *Herd Book of Hereford Cattle* in the 1995 volume. Les Cook, one of the first breeders active in the preservation of "Traditional" Herefords, remembers "calling

them something like 'pure English'" in the 1970s and 1980s—*not*, he is quick to point out, "with a desire to put anyone else's cattle down, it's just I knew the type of pedigree I was looking for." Interview with Les Cook, 4 January 2010.

30. Ritvo, "Race, Breed, and Myths of Origin."

31. The United Kingdom's 1908 livestock census "demonstrated the overwhelming numerical superiority of the dual-purpose Shorthorn." The breed constituted 64 percent of the national herd, estimated at 4.5 million cattle. Grundy, "Hereford Bull," 80.

32. Interview with Les Cook, 4 January 2010; "Llandinabo Farms Home Page," https://web.archive.org/web/20110830095004/http://www.llandinabofarms.co.uk/home.asp; "Traditional Hereford Beef," https://web.archive.org/web/20120328111549/http://www.traditionalherefords.org/hereford_beef.html. Emphasis original.

33. Ibid.

34. The controversy over imported Canadian Herefords was only slightly later than several waves of postcolonial human immigration to Great Britain, and if the commentary surrounding the "Traditional" Hereford was less explicit than Bailey's remarks in the 1880s, the implied parallels between human and animal were no less salient.

35. "Naval and Submarine Exhibit," *British Trade Journal*, 1 May 1882. Derbyshire Records Office, D1333 Z/Z 2.

36. Brown, *Sheep Farming*, 29.

37. Belich, *Replenishing the Earth*, 437–51.

38. "Sir Alfred Haslam, KT., J.P.: A Sketch of His Career," in *The Queen's State Visit to Derby May 21st, 1891*, 144. DRO, D1333 Z/Z 8.

39. Ibid.

40. Stokes, "Contesting Resources," esp. 48.

41. The degree to which British imperialism can be said to have been reciprocal is a subject of historiographical debate. For an overview, see Andrew S. Thompson, *Empire Strikes Back?*. For other facets, such as domestic consumption, identity, and popular culture, see Catherine Hall, *Civilising Subjects*; MacKenzie, *Imperialism and Popular Culture*; Hall and Rose, *At Home with the Empire*; Wilson, *New Imperial History*.

42. For post–World War II British immigration, particularly from the colonies and former colonies, see Holmes, *John Bull's Island*, esp. chap. 5; Hansen, *Citizenship and Immigration*; Karatani, *Defining British Citizenship*; Ryan and Webster, *Gendering Migration*.

43. For some of the cultural effects of colonial and postcolonial immigration, see Sauerberg, *Intercultural Voices*; and Dawson, *Mongrel Nation*.

44. Duckham, "History, Progress, and Comparative Merits," 10. Paper originally given at the Royal Agricultural College at Cirencester, 4 December 1863.

45. J. C. Hindson, "Questions on Trust Policy," *Ark* 1 (December 1974): 18.

Bibliography

Primary Sources

Unpublished Archival Sources

Derbyshire Record Office
   Haslam Foundry and Engineering Company Limited Papers, 1882–1912, GB 0026 D1333.
   Haslam's Patent Dry Air Refrigerators. Derbyshire Records Office, D1333 Z/Z 5.
   "Naval and Submarine Exhibit." *British Trade Journal*, 1 May 1882. Derbyshire Record Office, D1333 Z/Z 2.
   "Sir Alfred S. Haslam, KT., J.P.: A Sketch of His Career." In *The Queen's State Visit to Derby May 21st, 1891*. Derby U.K.: W. Hobson, 1891. Derbyshire Record Office D1333 Z/Z 8.
Hereford Cattle Association
   Export Ledger, 1890–1901.
   Export Ledger, 1901–1915.
Herefordshire Archive and Record Centre
   Hampton Court Estate
      Hereford Herd Book Society, A63/IV/46.
      John Hungerford Arkwright, correspondence, A63/IV/21; A63/IV/42.
      Unnamed subcollection, A63/III/65.
National Archives of Scotland
   Davidson, William Soltau. *The Establishment of the Frozen-Meat Trade, of the Dairying System, and of the Corriedale Breed of Sheep in New Zealand*. Edinburgh: New Zealand and Australian Land Company, 1918. GD435/614/6.

Published Periodicals

DAILY AND WEEKLY NEWSPAPERS
The Brisbane Courier (Queensland)
The Capricornian (Rockhampton, Queensland)
Graphic of Australia (Melbourne, Victoria)
Illustrated Sydney News (New South Wales)
Launceston Examiner (Tasmania)
The Maitland Mercury & Hunter River General Advertiser (New South Wales)
The Mercury (Hobart, Tasmania)
North Otago Times (Otago, New Zealand)
Otago Daily Times (Otago, New Zealand)
South Australian Register (Adelaide, South Australia)

Star (Christchurch, New Zealand)
Timaru Herald (Canterbury, New Zealand)

OTHER PERIODICALS
Agricultural Magazine, or, Farmers' Monthly Journal of Husbandry and Rural Affairs
All the Year Round
American Museum, or, Repository of Ancient & Modern Fugitive Pieces, &c. Prose & Poetical
Annals of Agriculture
The Ark
Australasian Pastoralists' Review
Chambers's Journal of Popular Literature, Science and Arts
Commercial and Agricultural Magazine
Country Life
European Magazine
Farmer's Magazine
Good Words
Journal of the Royal Agricultural Society of England
Leisure Hour
Livestock Journal and Fancier's Gazette
Monthly Miscellany
The New Review
New York Magazine
New Zealand Country Journal
New Zealand Farmer, Bee, and Poultry Journal
Quarterly Review
Reports of the Papers of the Lincolnshire Art and Archaeological Society
Saturday Review of Politics, Literature, Science and Art
Scots Magazine

Online Resources

Darwin, Charles. "Darwin to T. C. Eyton." *Darwin Correspondence Database*, entry 1946 (27 August 1856), http;//www.darwinproject.ac.uk/entry-1946, accessed on 12 January 2017.
———. "Darwin to Eyton." *Darwin Correspondence Database*, entry 1948 (31 August 1856), http;//www.darwinproject.ac.uk/entry-1948, accessed on 12 January 2017.
"Llandinabo Farms Home Page." *Internet Archive* (30 August 2011), https://web.archive.org/web/20110830095004/http://www.llandinabofarms.co.uk/home.asp, accessed on 12 January 2017.
Rare Breeds Survival Trust. "RBST Watchlist 2016." *Rare Breeds Survival Trust* (2016), http://www.rbst.org.uk/%252FOur-Work%252FResource-Library%252FWatchlists%252FWatchlist-2016, accessed on 12 January 2017.
———. "RBST Fact Sheet—Soay." *Rare Breeds Survival Trust*, http://www.rbst.org.uk/Rare-and-Native-Breeds/Sheep/Soay, accessed on 12 January 2017.

———. "Guidelines for Acceptance onto the Rare Breeds Survival Trust Watchlist." *Rare Breeds Survival Trust*, http://www.rbst.org.uk/Our-Work/Watchlist/About-the-Watchlist, accessed on 12 January 2017.

"Traditional Hereford Beef." *Internet Archive* (28 December 2011), https://web.archive.org/web/20120328111549/http://www.traditionalherefords.org/hereford_beef.html, accessed on 12 January 2017.

Oral History Interviews

Barnett, Philip. Oral History Interview by Rebecca Woods. Akitoa, New Zealand, 24 June 2010.

Cook, Les. Oral History Interview by Rebecca Woods. 4 January 2010.

Published Primary Sources

American Hereford Record. Vol. 1. Beecher, Ill.: Breeders' Live-Stock Association, 1880.

Archer, A. H., and James Sinclair. *Domestic Breeds and Their Treatment*. New ed. London: Vinton, 1896.

———. *Domestic Breeds and Their Treatment*. London: Vinton, 1896.

Armstrong, Albert Stapleton, and George Ord Campbell. *Australian Sheep Husbandry: A Handbook of the Breeding and Treatment of Sheep, and Station Management, with Concise Instructions for Tank and Well-Sinking, Fencing, Dam-Making, &c.* Melbourne: George Roberston, 1882.

Armstrong, Arthur Henry, Ross Grant, John S. Hogg, David Jones, W. H. Medcalf, Herbert Watkins-Pitchford, R. Ramsay, Joseph Raymond, Gustavo Rey-Alverz, Juan E. Richelet, J. Watson, and Thomas Dunlop Young. *The Frozen and Chilled Meat Trade: A Practical Treatise by Specialists in the Meat Trade*. 2 vols. London: Gresham, 1929.

Banks, Joseph. "Address to the Members." In *The Second Report of the Merino Society*, 5–8. London: Evans & Ruffy, 1812.

———. *A Project for Extending the Breed of Fine-Wooled Spanish Sheep, Now in the Possession of His Majesty, into All Parts of Great Britain, Where the Growth of Fine Clothing Wools Is Found to Be Profitable*. London: W. Bulmer, 1804.

———. *Some Circumstances Relative to Merino Sheep: Chiefly Collected from the Spanish Shepherds, Who Attended Those of the Flock Paular, Lately Presented to His Majesty by the Government of Spain; with Particulars Respecting that Great National Acquisition; and Also Respecting the Sheep of the Flock of Negrete*. London: W. Bulmer, 1809.

Bradford, William. *Bradford's History of Plymouth Plantation, 1606–1646*. New York: Barnes & Noble, 1982.

Broodbank, Joseph G. "The Development of Refrigerated Accommodation in British Ports." In *Proceedings of the Fourth International Congress of Refrigeration, Held under the Auspices of the International Institute of Refrigeration, 16–21 June, 1924*, vol. 2, 1705–20. London: International Refrigerating Congress Movement, 1924.

Brown, William. *British Sheep Farming*. Edinburgh: Adam and Charles Black, 1870.
Bucke, Thomas George. "Observations on the Quality of the Mutton." In *Second Report of the Merino Society*, 10–14. London: Evans & Ruffy, 1812.
———. "Report." In *Third Report of the Merino Society*, 5–89. London: Evans & Ruffy, 1813.
Butler, Samuel. *A First Year in Canterbury Settlement with Other Early Essays*. Edited by R. A. Streatfeild. London: A. C. Fifield, 1914.
Coates, George. *The General Short-Horned Herd-Book: Containing the Pedigrees of Short-horned Bulls, Cows, etc. of the Improved Durham Breed, from the Earliest Account to the Year 1822*. Otley, U.K.: W. Walker, 1822.
The Corriedale, New Zealand's Own Breed: History and Development. Christchurch, N.Z.: The Corriedale Sheep Society, 1936.
Coventry, Andrew. *Remarks on Live Stock and Relative Subjects*. Edinburgh: Archibald Constable, 1806.
Critchell, James Troubridge, and Joseph Raymond. *A History of the Frozen Meat Trade: An Account of the Development and Present Day Methods of Preparation, Transport, and Marketing of Frozen and Chilled Meats*. London: Constable, 1912.
Culley, George. *Observations on Live Stock: Containing Hints for Choosing and Improving the Best Breeds of the Most Useful Kinds of Domestic Animals*. London: G. Wilkie & J. Robinson, 1807.
———. *Observations on Live Stock, Containing Hints for Choosing and Improving the Best Breeds of the Most Useful Kinds of Domestic Animals*. London: G. G. & J. Robinson, 1786.
Cyclopedia of New Zealand. Wellington Provincial District. Wellington, N.Z.: Cyclopedia Company, 1897.
Darwin, Charles. *On the Origin of Species by Means of Natural Selection, or, the Preservation of Favored Races in the Struggle for Life*. London: John Murray, 1859.
———. *The Variation of Animals and Plants under Domestication*. Vol. 1. Baltimore: Johns Hopkins University Press, 1998.
Davidson, William Soltau. *William Soltau Davidson, 1846–1924: A Sketch of His Life Covering a Period of Fifty-two Years, 1864–1916, in the Employment of the New Zealand and Australian Land Company Limited*. Edinburgh: Oliver and Boyd, 1930.
de Lasteyrie, Charles. *An Account of the Introduction of Merino Sheep into the Different States of Europe, and at the Cape of Good Hope*. Translated by Benjamin Thompson. London: John Harding, 1810.
Duckham, Thomas. *Eyton's Herd Book of Hereford Cattle*, vol. 3. Hereford: William Phillips, 1858.
———. "A Lecture on the Breeding and Management of Hereford Cattle." In *Eyton's Herd Book of Hereford Cattle*, vol. 7, 3–15. Hereford: Longman, 1869.
———. "A Lecture on the History, Progress, and Comparative Merits of the Hereford Breed of Cattle." In *Eyton's Herd Book of Hereford Cattle*, vol. 6, 3–32. Hereford: Longman, 1868.
Duncumb, John. *General View of the Agriculture of the County of Hereford; Drawn up for the Consideration of the Board of Agriculture and Internal Improvement*. London: Bulmer, 1805.

Evans, B. L. *Agricultural and Pastoral Statistics of New Zealand, 1861–1954*. Wellington, N.Z.: R. E. Owen, Government Printer, 1956.
"Extracts from the Minutes of the Smithfield Club from 1798 to 1900." In *History of the Smithfield Club, from 1798 to 1900*, by Edwin James Powell, 19–35. London: Smithfield Club, 1900.
Eyton, Thomas Campbell. *A Catalogue of British Birds*. London: Longman, Rees, Brown, and Green, 1836.
———. *The Herd Book of Hereford Cattle*. 2 vols. London: Longman, 1846–1853.
———. *A History of the Rarer British Birds*. London: Longman, Rees, Brown, and Green, 1836.
———. "Preface." In *The Herd Book of Hereford Cattle*, vol. 2, iv. London: Longman, 1853.
Fowler, J. Kersley. *Records of Old Times: Historical, Social, Political, Sporting and Agricultural*. London: Chatto & Windus, 1898.
Gooch, William. *General View of the Agriculture of the County of Cambridge*. London: Richard Phillips, 1811.
Graham, John Ryrie. *A Treatise on the Australian Merino*. Melbourne: Clarson, Massina, 1870.
Grant, Ross. "The Australian Meat Industry." In *The Frozen and Chilled Meat Trade: A Practical Treatise by Specialists in the Trade*, edited by Arthur Henry Armstrong et al., vol. 1, 31–100. London: Gresham, 1929.
Halifax. "Foreword." In *Britain Can Breed It*, 2nd ed., 4–5. London: Farmer and Stockbreeder, 1949.
Hall, George Webb. "Observations on the Growth and Management of Merino Wool." In *The Second Report of the Merino Society*, 36–65. London: Evans & Ruffy, 1812.
Hereford Herd Book Society. *Herd Book of Hereford Cattle*. Vols. 10 and 11. Hereford: Hereford Herd Book Society, 1879–1880.
Holford, G. H. *The Corriedale: New Zealand's Own Breed*. Christchurch, N.Z.: Corriedale Sheep Society, 1928.
———. *The Corriedale: New Zealand's Contribution to the Sheep World*. Christchurch, N.Z.: The Corriedale Sheep Society, 1924.
House of Lords. *Report from the Select Committee on Marking of Foreign Meat, &c.; Together with the Preceedings of the Committee, Minutes of Evidence, and Appendix*. Select Committee, House of Lords. Great Britain, Parliament, August 24, 1893.
Hunt, Charles Henry. *A Practical Treatise on the Merino and Anglo-Merino Breeds of Sheep: In Which the Advantages to the Farmer and Grazier, Peculiar to These Breeds, Are Clearly Demonstrated*. London: Printed for W. P. Piercy, 1809.
Hunt, John. *Agricultural Memoirs; or History of the Dishley System. In Answer to Sir John Saunders Sebright, Bart. M.P.* Nottingham: Printed for the author by H. Barnett, 1812.
Jefferson, Thomas. *Notes on the State of Virginia*. New York: Penguin Books, 1999.
Jones, David. "New Zealand Trade." In *The Frozen and Chilled Meat Trade: A Practical Treatise by Specialists in the Trade*, edited by Arthur Henry Armstrong et al., vol. 1, 101–58. London: Gresham, 1929.

Knight, Thomas Andrew. "An Account of some Experiments of the Fecundation of Vegetables, in a Letter from Thomas Andrew Knight, Esq., to the Right Hon. Sir Joseph Banks, K. B. P. R. S." *Philosophical Transactions of the Royal Society of London* 89 (1799): 195–204.

———. "On the Hereditary Instinctive Propensities of Animals." *Philosophical Transactions of the Royal Society of London* 127 (1837): 365–69.

Lawrence, John. *A General Treatise on Cattle, the Ox, the Sheep, and the Swine.* London: H. D. Symonds, 1805.

Little, James. *The Story of the Corriedale: Also a Few Suggestions as to the Possible Cause of Black Sheep.* Christchurch, N.Z.: Willis and Aitken, 1917.

Little, John. *Practical Observations on the Improvement and Management of Mountain Sheep, and Sheep Farms.* Edinburgh: Printed by John Moir for Macredie, 1815.

Low, David. *The Breeds of the Domestic Animals of the British Islands.* London: Longman, Brown, Green & Longmans, 1842.

———. *On the Domestic Animals of the British Islands: Comprehending the Natural and Economical History of Species and Varieties; the Description of the Properties of External Form; and Observations on the Principles and Practice of Breeding.* London: Longman, Brown, Green & Longmans, 1845.

MacDonald, James. *Food from the Far West.* London: W. P. Nimmo, 1878.

MacDonald, James, and James Sinclair. *History of Hereford Cattle.* London: Vinton, 1909.

Marshall, William. *The Rural Economy of Gloucestershire; Including Its Dairy: Together with the Dairy Management of North Wiltshire; and the Management of Orchards and Fruit Liquor, in Herefordshire.* Vol. 2, 2nd ed. London: G. Nicol, 1796.

Marshall, William Humphrey. *The Rural Economy of the West of England: Including Minutes of Practice, in That Department.* Vol. 1, 2nd ed. London: G. & W. Nicol, 1805.

———. *The Rural Economy of Yorkshire, Comprizing the Management of Landed Estates, and the Present Practice of Husbandry in the Agricultural Districts of the Country.* Vol. 2, 2nd ed. London: G. Nicol, 1796.

Montalivet. "Report of the Minister of the Interior." In *The First Report of the Merino Society.* London: Evans & Ruffy, 1811.

New Zealand Herd Book (of Breeds of Cattle other than Short-Horns) Embracing Herefords, Ayrshires, Polled Angus, Channel Islands', Devons and Dutch Friesian. Vol. 1. Christchurch, N.Z.: Canterbury Agricultural and Pastoral Association, 1886.

Parkinson, Richard. *Treatise on the Breeding and Management of Live Stock, in Which the Principles and Proceedings of the New School of Breeders Are Fully and Experimentally Discussed.* Vol. 1. London: Cadell and Davies, 1810.

Parry, Caleb Hillier. *Facts and Observations Tending to Shew the Practicability and Advantage, to the Individual and the Nation, of Producing in the British Isles Clothing Wool, Equal to That of Spain: Together with Some Hints towards the Management of Fine-Woolled Sheep.* London: Cadell and Davies, 1800.

Pearce, William. *General View of the County of Berkshire.* London: W. Bulmer, 1794.

Powell, Edwin James. *History of the Smithfield Club from 1798 to 1900.* London: Smithfield Club, 1902.

Proceedings of the Fourth International Congress of Refrigeration, Held under the Auspices of the International Institute of Refrigeration, 16–21 June, 1924. 2 vols. London: International Refrigerating Congress Movement, 1924.

Ramsay, R. "The Rise of the World's Refrigerated Meat Traffic, and Its Effect on the Resources of the Various Countries of Meat Supply." In *Proceedings of the Fourth International Congress of Refrigeration, Held under the Auspices of the International Institute of Refrigeration, 16–21 June, 1924*, vol. 2, 1720–33. London: International Refrigerating Congress Movement, 1924.

———. "The World's Frozen and Chilled Meat Trade." In *The Frozen and Chilled Meat Trade: A Practical Treatise by Specialists in the Trade*, edited by Arthur Henry Armstrong et al., vol. 1, 3–30. London: Gresham, 1929.

Rudge, Thomas. *General View of the Agriculture of the County of Gloucester*. London: Richard Phillips, 1807.

Sebright, John Saunders. *The Art of Improving the Breeds of Domestic Animals: In a Letter Addressed to the Right Hon. Sir Joseph Banks, K. B.* London: Printed for J. Harding, 1809.

Sinclair, John. *Observations on the Different Breeds of Sheep, and the State of Sheep Farming in Some of the Principal Counties of England*. Edinburgh: W. Smellie, 1792.

Somerville, John Southey. *Facts and Observations Relative to Sheep, Wool, Ploughs and Oxen: In Which the Importance of Improving the Short-Wooled Breeds of Sheep, by a Mixture of the Merino Blood, Is Demonstrated from Actual Practice*. 3rd ed. London: Printed for J. Harding, 1809.

———. *Lord Somerville's Address to the Board of Agriculture: On the Subject of Sheep and Wool, on the 14th of May 1799*. Sussex: John Lord Sheffield for the Board of Agriculture, 1799.

Statistical Abstract for the British Empire in Each Year from 1889 to 1903. London: His Majesty's Stationery Office, 1905.

Statistical Abstract for the Several Colonial and Other Possessions of the United Kingdom in Each Year from 1876 to 1890. Vol. 28. London: Her Majesty's Stationery Office, 1891.

Thompson, Benjamin. "Appendix: Letter from B. Thompson, Esq., to Sir Joseph Banks." In *The Second Report of the Merino Society*, 117–61. London: Evans & Ruffy, 1812.

———. "Preface to First Report." In *The First Report of the Merino Society*, iii–iv. London: Evans & Ruffy, 1811.

Wallis-Tayler, A. J. *Refrigeration, Cold Storage and Ice-Making: A Practical Treatise on the Art and Science of Refrigeration*. 5th ed. London: Crosby Lockwood and Son, 1917.

Waters, Sydney D. *Clipper Ship to Motor Liner: The Story of the New Zealand Shipping Company 1873–1939*. London: The New Zealand Shipping Company, 1939.

Welles, E. F. *Remarks and Suggestions on the Form of Cattle, with Illustrations Indicative of the True and the Defective*. Hereford: J. Head, n.d.

Worgan, George B. *General View of the Agriculture of the County of Cornwall*. London: B. McMillan, 1807.

Youatt, William. *Cattle: Their Breeds, Management, and Diseases*. London: Baldwin and Craddock, 1834.

———. *The Complete Grazier: Or, Farmer's and Cattle Breeder's and Dealer's Assistant.* 6th ed. London: Baldwin and Craddock, 1833.

———. *Sheep: Their Breeds, Management, and Diseases. To Which Is Added the Mountain Shepherd's Manual.* London: Baldwin and Craddock, 1837.

Young, Arthur. *A Six Weeks' Tour through the Southern Counties of England and Wales.* London: W. Strahan and W. Nicoll, 1769.

Secondary Sources

Anderson, Virginia DeJohn. *Creatures of Empire: How Domestic Animals Transformed Early America.* Oxford: Oxford University Press, 2004.

Anderson, Warwick. *The Cultivation of Whiteness: Science, Health, and Racial Destiny in Australia.* Durham, N.C.: Duke University Press, 2006.

———. "Disease, Race, and Empire." *Bulletin of the History of Medicine* 70, no. 1 (1996): 62–67.

Beattie, James, Emily O'Gorman, and Matthew Henry, eds. *Climate, Science, and Colonization: Histories from Australia and New Zealand.* New York: Palgrave Macmillan, 2014.

Belich, James. *Replenishing the Earth: The Settler Revolution and the Rise of the Anglo-World, 1783–1939.* Oxford: Oxford University Press, 2009.

Breen, Benjamin. "'The Elks Are Our Horses': Animals and Domestication in the New France Borderlands." *Journal of Early American History* 3 (2013): 181–206.

Breen, T. H. "An Empire of Goods: The Anglicization of Colonial America, 1690–1776." *Journal of British Studies* 25, no. 4 (1 October 1986): 467–99.

Broers, Michael. *Europe under Napoleon, 1799–1815.* New York: Arnold, 1996.

Brooking, Tom, and Eric Pawson, eds. "The Contours of Transformation." In *Seeds of Empire: The Environmental Transformation of New Zealand*, edited by Tom Brooking and Eric Pawson, 13–33. London: I. B. Tauris, 2011.

———. *Seeds of Empire: The Environmental Transformation of New Zealand.* London: I. B. Tauris, 2011.

Burton, Antoinette, and Dane Kennedy, eds. *How Empire Shaped Us.* London: Bloomsbury Academic, 2016.

Campbell, R. N. "St Kilda and Its Sheep." In *Island Survivors: The Ecology of the Soay Sheep of St Kilda*, edited by P. A. Jewell, C. Milner, and J. Morton Boyd, 8–35. London: Athlone Press of the University of London, 1974.

Carter, H. B. *His Majesty's Spanish Flock; Sir Joseph Banks and the Merinos of George III of England.* Sydney: Angus & Robertson, 1964.

Cassidy, Rebecca. *The Sport of Kings: Kinship, Class, and Thoroughbred Breeding in Newmarket.* Cambridge: Cambridge University Press, 2002.

Chaplin, Joyce E. "Creoles in British America: From Denial to Acceptance." In *Creolization: History, Ethnography, Theory*, edited by Charles Stewart, 46–65. Walnut Creek, Calif.: Left Coast Press, 2007.

———. *Subject Matter: Technology, the Body, and Science on the Anglo-American Frontier, 1500–1676.* Cambridge, Mass.: Harvard University Press, 2001.

Clutton-Brock, Juliet. *A Natural History of Domesticated Mammals*. Cambridge: Cambridge University Press, 1987.

———. *The Walking Larder: Patterns of Domestication, Pastoralism, and Predation*. London: Unwin Hyman, 1989.

Clutton-Brock, Juliet, and Steven G. J. Hall. *Two Hundred Years of British Farm Livestock*. London: Natural History Museum, 1995.

Clutton-Brock, T. H., J. M. Pemberton, T. Coulson, I. R. Stevenson, and A. D. C. MacColl. "The Sheep of St Kilda." In *Soay Sheep: Dynamics and Selection in an Island Population*, edited by T. H. Clutton-Brock and Josephine Pemberton, 17–51. Cambridge: Cambridge University Press, 2004.

Colley, Linda. *Britons: Forging the Nation, 1707–1837*. New Haven, Conn.: Yale University Press, 1992.

Collins, E. J. T. "Food Supplies and Food Policy." In *The Agrarian History of England and Wales*, edited by E. J. T. Collins and Joan Thirsk, vol. 7, pt. 1, 33–71. Cambridge: Cambridge University Press, 2000.

———. "Rural and Agricultural Change." In *The Agrarian History of England and Wales*, edited by E. J. T. Collins and Joan Thirsk, vol. 7, pt. 1, 72–223. Cambridge: Cambridge University Press, 2000.

Cook, Harold. *Matters of Exchange: Commerce, Medicine, and Science in the Dutch Golden Age*. New Haven, Conn.: Yale University Press, 2007.

Cronon, William. *Changes in the Land: Indians, Colonists, and the Ecology of New England*. New York: Hill and Wang, 1983.

———. *Nature's Metropolis: Chicago and the Great West*. New York: W. W. Norton, 1991.

Crosby, Alfred W. *The Columbian Exchange: Biological and Cultural Consequences of 1492*. Westport, Conn.: Greenwood, 1972.

———. *Ecological Imperialism: The Biological Expansion of Europe, 900–1900*. Cambridge: Cambridge University Press, 1994.

———. "Virgin Soil Epidemics." In *American Environmental History*, edited by Louis S. Warren, 50–62. Oxford: Blackwell, 2003.

Dawson, Ashley. *Mongrel Nation: Diasporic Culture and the Making of Postcolonial Britain*. Ann Arbor: University of Michigan Press, 2007.

Derry, Margaret Elsinor. *Bred for Perfection: Shorthorn Cattle, Collies, and Arabian Horses Since 1800*. Baltimore: Johns Hopkins University Press, 2003.

———. *Masterminding Nature: The Breeding of Animals 1750–2010*. Toronto: University of Toronto Press, Scholarly Publishing Division, 2015.

———. *Ontario's Cattle Kingdom: Purebred Breeders and Their World, 1870–1920*. Toronto: University of Toronto Press, 2001.

Dunlap, Thomas. *Nature and the English Diaspora: Environment and History in the United States, Canada, Australia, and New Zealand*. New York: Cambridge University Press, 1999.

Farland, Maria. "Modernist Versions of Pastoral: Poetic Inspiration, Scientific Expertise, and the 'Degenerate' Farmer." *American Literary History* 19, no. 4 (2007): 905–36.

Flannery, Tim F. *The Future Eaters: An Ecological History of the Australasian Lands and People*. Chatswood, N.S.W.: Reed, 1994.

Franklin, Sarah. *Dolly Mixtures: The Remaking of Genealogy*. Durham, N.C.: Duke University Press, 2007.

Fraser Darling, Frank. "Foreword." In *Island Survivors: The Ecology of the Soay Sheep of St Kilda*, edited by P. A. Jewell, C. Milner, and J. Morton Boyd, x–xii. London: Athlone Press of the University of London, 1974.

Freeman, Sarah. *Mutton and Oysters: The Victorians and Their Food*. London: V. Gollancz, 1989.

Fussell, G. E. "Four Centuries of Lincolnshire Farming." *Reports of Papers of the Lincolnshire Art and Archaeological Society* 4 (1952): 4–37.

Grasseni, Christina. *Skilled Visions: Between Apprenticeship and Standards*. New York: Berghahn Books, 2007.

Grundy, Joan E. "The Hereford Bull: His Contribution to New World and Domestic Beef Supplies." *Agricultural History Review* 50, no. 1 (2002): 69–88.

Hall, Catherine. *Civilising Subjects: Colony and Metropole in the English Imagination, 1830–1867*. Chicago: Chicago University Press, 2002.

———, ed. *Cultures of Empire: A Reader: Colonizers in Britain and the Empire in the Nineteenth and Twentieth Centuries*. Manchester: Manchester University Press, 2000.

Hall, Catherine, and Sonya O. Rose, eds. *At Home with the Empire: Metropolitan Culture and the Imperial World*. Cambridge: Cambridge University Press, 2006.

Hansen, Randall. *Citizenship and Immigration in Post-War Britain: The Institutional Origins of a Multicultural Nation*. Oxford: Oxford University Press, 2000.

Harley, C. Knick. "Steers Afloat: The North Atlantic Meat Trade, Liner Predominance, and Freight Rates, 1870–1913." *Journal of Economic History* 68, no. 4 (December 2008): 1028–58.

Harman, Mary. *An Isle Called Hirte: History and Culture of the St Kildans to 1930*. Waternish, Isle of Skye: Maclean Press, 1997.

Harrison, Mark. *Climates & Constitutions: Health, Race, Environment and British Imperialism in India, 1600–1850*. New Delhi: Oxford University Press, 1999.

Heath-Agnew, E. *A History of Hereford Cattle: And Their Breeders*. London: Duckworth, 1983.

Higgins, David M. "'Mutton Dressed as Lamb?' The Misrepresentation of Australian and New Zealand Meat in the British Market, c. 1890–1914." *Australasian Economic History Review* 44, no. 2 (July 2004): 161–84.

Holmes, Colin. *John Bull's Island: Immigration and British Society, 1871–1971*. Basingstoke: Macmillan, 1988.

Isenberg, Andrew C. *The Destruction of the Bison: An Environmental History 1750–1920*. Cambridge: Cambridge University Press, 2000.

Jacoby, Karl. *Crimes against Nature: Squatters, Poachers, Thieves, and the Hidden History of American Conservation*. Berkeley: University of California Press, 2001.

Jewell, P. A., C. Milner, and J. Morton Boyd, eds. *Island Survivors: The Ecology of the Soay Sheep of St Kilda*. London: Athlone Press of the University of London, 1974.

Johnson, Walter. *Soul by Soul: Life inside the Antebellum Slave Market*. Cambridge, Mass.: Harvard University Press, 1999.

Jones, David S. *Rationalizing Epidemics: Meanings and Uses of American Indian Mortality since 1600.* Cambridge, Mass.: Harvard University Press, 2004.

———. "Virgin Soils Revisited." *William and Mary Quarterly* 60, no. 4 (1 October 2003): 703–42.

Jonsson, Fredrik Albritton. *Enlightenment's Frontier: The Scottish Highlands and the Origins of Environmentalism.* New Haven, Conn.: Yale University Press, 2013.

Jordan, Terry G. *North American Cattle-Ranching Frontiers: Origins, Diffusion, and Differentiation.* Albuquerque: University of New Mexico Press, 1993.

Karatani, Rieko. *Defining British Citizenship: Empire, Commonwealth, and Modern Britain.* London: Frank Cass, 2003.

Kennedy, Dane. "The Perils of the Midday Sun: Climatic Anxieties in the Colonial Tropics." In *Imperialism and the Natural World*, edited by John D. MacKenzie, 118–40. Manchester: Manchester University Press, 1990.

Kupperman, Karen Ordahl. "Fear of Hot Climates in the Anglo-American Colonial Experience." *William and Mary Quarterly* 41, no. 2 (1984): 213–40.

Langford, Paul. "The Eighteenth Century." In *The Oxford History of Britain*, edited by Kenneth O. Morgan, 399–469. Oxford: Oxford University Press, 2010.

Lee, Paula Young, ed. *Meat, Modernity, and the Rise of the Slaughterhouse.* Durham: University of New Hampshire Press, 2008.

MacKenzie, John M., ed. *Imperialism and Popular Culture.* Manchester: Manchester University Press, 1986.

Magee, Gary Bryan, and Andrew S. Thompson. *Empire and Globalisation : Networks of People, Goods and Capital in the British World, c.1850–1914.* New York: Cambridge University Press, 2010.

McCook, Stuart. "The Neo-Columbian Exchange: The Second Conquest of the Greater Caribbean, 1720–1930." *Latin American Research Review* 46 (2011): 11–31.

McLintock, A. H. *An Encyclopedia of New Zealand.* 3 vols. Wellington, N.Z.: R. E. Owen, Government Printer, 1966.

McNeill, John Robert. *Mosquito Empires: Ecology and War in the Greater Caribbean, 1620–1914.* New York: Cambridge University Press, 2010.

McNeill, William H. *Plagues and Peoples.* New York: Doubleday, 1989.

Melville, Elinor G. K. *A Plague of Sheep: Environmental Consequences of the Conquest of Mexico.* Cambridge: Cambridge University Press, 1994.

Metcalfe, Robyn S. *Meat, Commerce and the City: The London Food Market, 1800–1855.* London: Pickering and Chatto, 2012.

Mintz, Sidney. *Sweetness and Power: The Place of Sugar in Modern History.* New York: Viking, 1985.

Mitchell, B. R. *Abstract of British Historical Statistics.* Cambridge: Cambridge University Press, 1962.

Morgan, Raine, and G. E. Mingay. "Root Crops." In *Agrarian History of England and Wales*, edited by G. E. Mingay and Joan Thirsk, vol. 6, 285–314. Cambridge: Cambridge University Press, 1989.

Morton Boyd, J. "Introduction." In *Island Survivors: The Ecology of the Soay Sheep of St Kilda*, edited by P. A. Jewell, C. Milner, and J. Morton Boyd, 1–7. London: Athlone Press of the University of London, 1974.

Morton Boyd, J., and P. A. Jewell. "The Soay Sheep and Their Environment: A Synthesis." In *Island Survivors: The Ecology of the Soay Sheep of St Kilda*, edited by P. A. Jewell, C. Milner, and J. Morton Boyd, 360–73. London: Athlone Press of the University of London, 1974.

Muir, Cameron. *The Broken Promise of Agricultural Progress: An Environmental History*. Oxford: Routledge, 2014.

Müller-Wille, Staffan, and Hans-Jörg Rheinberger. "Heredity—The Formation of an Epistemic Space." In *Heredity Produced: At the Crossroads of Biology, Politics, and Culture, 1500–1870*, edited by Staffan Müller-Wille and Hans-Jörg Rheinberger, 3–33. Cambridge, Mass.: MIT Press, 2007.

Otter, Chris. "Civilizing Slaughter: The Development of the British Public Abattoir, 1850–1910." In *Meat, Modernity, and the Rise of the Slaughterhouse*, edited by Paula Young Lee, 89–106. Durham: University of New Hampshire Press, 2008.

Pachirat, Timothy. *Every Twelve Seconds : Industrialized Slaughter and the Politics of Sight*. New Haven, Conn.: Yale University Press, 2011.

Pawley, Emily. "The Point of Perfection: Cattle Portraiture, Bloodlines, and the Meaning of Breeding, 1760–1860." *Journal of the Early American Republic* 36, no. 1 (Spring 2016): 37–72.

Pawson, Eric, and Tom Brooking. "Introduction." In *Seeds of Empire: The Environmental Transformation of New Zealand*, edited by Tom Brooking and Eric Pawson, 1–12. London: I. B. Tauris, 2011.

Peden, Robert L. "Pastoralism and the Transformation of the Open Grasslands." In *Seeds of Empire: The Environmental Transformation of New Zealand*, edited by Tom Brooking and Eric Pawson, 73–93. London: I. B. Tauris, 2011.

Perren, Richard. *The Meat Trade in Britain, 1840–1914*. London: Routledge and Kegan Paul, 1978.

Pomeranz, Kenneth. *The Great Divergence: China, Europe, and the Making of the Modern World Economy*. Princeton, N.J.: Princeton University Press, 2000.

Prince, Hugh C. "The Changing Rural Landscape, 1750–1850." In *Agrarian History of England and Wales*, edited by G. E. Mingay and Joan Thirsk, vol. 6, 7–83. Cambridge: Cambridge University Press, 1989.

Richards, Eric. *From Hirta to Port Phillip: The Story of the Ill-fated Emigration from St Kilda to Australia in 1852*. Ravenspoint, U.K.: Islands Book Trust, 2010.

Ritvo, Harriet. *The Animal Estate: The English and Other Creatures in the Victorian Age*. Cambridge, Mass.: Harvard University Press, 1987.

———. "Mad Cow Mysteries." In *Noble Cows and Hybrid Zebras: Essays on Animals and History*. Charlottesville: University of Virginia Press, 2010.

———. *The Platypus and the Mermaid and Other Figments of the Classifying Imagination*. Cambridge, Mass.: Harvard University Press, 1997.

———. "Possessing Mother Nature: Genetic Capital in Eighteenth-Century Britain." In *Early Modern Conceptions of Property*, edited by John Brewer and Susan Staves, 413–26. London: Routledge, 1995.

———. "Race, Breed, and Myths of Origin: Chillingham Cattle as Ancient Britons." In *Noble Cows and Hybrid Zebras: Essays on Animals and History*, 132–56. Charlottesville: University of Virginia Press, 2010.

Robin, Libby. *How a Continent Created a Nation.* Sydney: University of New South Wales Press, 2007.

Rogers, Ben. *Beef and Liberty.* London: Chatto & Windus, 2003.

Rosenberg, Gabriel N. "A Race Suicide among the Hogs: The Biopolitics of Pork Production in the United States, 1865-1930." *American Quarterly* 68, no. 1 (March 2016): 49–73.

Russell, Nicholas. *Like Engend'ring Like: Heredity and Animal Breeding in Early Modern England.* Cambridge: Cambridge University Press, 1986.

Ryan, Louise, and Wendy Webster. *Gendering Migration: Masculinity, Femininity and Ethnicity in Post-War Britain.* Aldershot, U.K.: Ashgate, 2008.

Saraiva, Tiago. *Fascist Pigs: Technoscientific Organisms and the History of Fascism.* Cambridge, Mass.: MIT Press, 2016.

Sauerberg, Lars Ole. *Intercultural Voices in Contemporary British Literature: The Implosion of Empire.* Houndmills, U.K.: Palgrave, 2001.

Sayer, Karen. "Animal Machines: The Public Response to Intensification in Great Britain, c. 1960–c. 1973." *Agricultural History* 87, no. 4 (Fall 2013): 473–501.

Sayre, Nathan Freeman. *Ranching, Endangered Species, and Urbanization in the Southwest: Species of Capital.* Tucson: University of Arizona Press, 2002.

Shapin, Steven. "'You Are What You Eat': Historical Changes in Ideas about Food and Identity." *Historical Research* 87, no. 237 (1 August 2014): 377–92.

Shukin, Nicole. *Animal Capital: Rendering Life in Biopolitical Times.* Minneapolis: University of Minnesota Press, 2009.

Sluyter, Andrew. "The Making of the Myth in Postcolonial Development: Material-Conceptual Landscape Transformation in Sixteenth-Century Veracruz." *Annals of the Association of American Geographers* 89, no. 3 (1999): 377–401.

Steel, Frances. *Oceania under Steam: Sea Transport and the Cultures of Colonialism, c. 1870–1914.* Manchester: Manchester University Press, 2012.

Stokes, Evelyn. "Contesting Resources: Maori, Pakeha, and a Tenurial Revolution." In *Environmental Histories of New Zealand,* edited by Eric Pawson and Tom Brooking, 35–51. Oxford: Oxford University Press, 2002.

Stringleman, Hugh, and Robert Peden. "Sheep Farming: The Establishment Phase." *Te Ara: the Encyclopedia of New Zealand* (13 July 2012), http://www.TeAra.govt.nz/en/sheep-farming/page-2. 5 January 2016.

Theunissen, Bert. "Breeding without Mendelism: Theory and Practice of Dairy Cattle Breeding in the Netherlands, 1900–1950." *Journal of the History of Biology* 41 (2008): 637–76.

Thompson, Andrew S. *The Empire Strikes Back? The Impact of Imperialism on Britain from the Mid-Nineteenth Century.* Harlow: Pearson Longman, 2005.

Trow-Smith, Robert. *A History of British Livestock Husbandry,* 2 vols. London: Routledge and Kegan Paul, 1957–1959.

Turner, Michael. *Enclosures in Britain 1750–1830.* London: Macmillan, 1984.

Valenze, Deborah M. *Milk : A Local and Global History.* New Haven, Conn.: Yale University Press, 2011.

Vernon, James. *Hunger: A Modern History.* Cambridge: Belknap Press of Harvard University Press, 2007.

Vialles, Noelie. *Animal to Edible*. Translated by J. A. Underwood. Cambridge: Cambridge University Press, 1994.

White, Richard. *Railroaded: The Transcontinentals and the Making of Modern America*. New York: W. W. Norton, 2011.

Wild, Trevor. *Village England: A Social History of the Countryside*. London: I. B. Tauris, 2004.

Wilson, Kathleen, ed. *A New Imperial History: Culture, Identity, and Modernity in Britain and the Empire, 1600–1850*. Cambridge: Cambridge University Press, 2004.

Wood, Roger J. "The Sheep Breeders' View of Heredity before and after 1800." In *Heredity Produced: At the Crossroads of Biology, Politics, and Culture, 1500–1870*, edited by Staffan Müller-Wille and Hans-Jörg Rheinberger, 229–49. Cambridge, Mass.: MIT Press, 2007.

Wood, Roger J., and Vítězslav Orel. *Genetic Prehistory in Selective Breeding: A Prelude to Mendel*. Oxford: Oxford University Press, 2001.

Woods, Abigail. "Breeding Cows, Maximising Milk: British Veterinarians and the Livestock Economy, 1930–50." In *Healing the Herds: Disease, Livestock Economies, and the Globalization of Veterinary Medicine*, edited by Karen Brown and Daniel Gilfoyle, 59–75. Athens: Ohio University Press, 2010.

Woods, Rebecca J. H. "Breed, Culture, and Economy: The New Zealand Frozen Meat Trade, 1880–1914." *Agricultural History Review* 60, no. 2 (2012): 288–308.

———. "From Colonial Animal to Imperial Edible: Building an Empire of Sheep in New Zealand, c. 1880–1900." *Comparative Studies of South Asia, Africa, and the Middle East* 35, no. 1 (2015): 117–36.

———. "Green Mountain Merinos: Sheep Breeding in Nineteenth-Century Vermont." *Vermont History* 85, 1 (2017): 1–19.

Worcester, Donald E. *The Texas Longhorn, Relic of the Past, Asset for the Future*. College Station: Texas A & M University Press, 1987.

Worster, Donald. *Dust Bowl: The Southern Plains in the 1930s*. New York: Oxford University Press, 1979.

Zilberstein, Anya. *A Temperate Empire: Climate Change and Settler Colonialism in Early North America*. New York: Oxford University Press, 2016.

Index

Note: Illustrations are indicated by page numbers in italics.

Aberdeen-Angus cattle, 207n85, 211n26
Acclimatization: crossbreeding and, 134; of Hereford cattle, 143–44; of humans, 12–13, 40; nativeness and, 12–13; of sheep, 33, 58; wool and, 69. *See also* Climate
"Agricola" (writer), 39, 87
Agricultural Gazette, 165
Agricultural improvement, 32; in Americas, 150–51, 159; breed identity and, 5; conservation and, 167; crossbreeding and, 45, 90; environment as limit to, 48–49; and Herefords vs. Shorthorns, 80, 85, 89, 193n149; inbreeding and, 99–100; meat and, 47, 72, 82–83, 87–88; merinos and, 65–66, 72–75; native breeds and, 25; of New Leicester Longwool, 41–42; and notion of "breed," 29–30; patriotism and, 16–17, 46, 63; type and, 41; wool and, 65, 130
Agricultural Magazine, 53, 59, 70, 72, 86–87, 90, 141
All the Year Round, 117, 119
Altar Valley, Utah, 10
American Hereford Cattle Breeders' Association, 161–62
American Hereford Record, 161, 209n159
Anderson, Virginia DeJohn, 9
"Anglo World," 16
Angus cattle, 4, 140
Annals of Agriculture, 49, 89
Anson, T. H., 130–31
Archer, A. H., 43
Archer, Robert, 155

Argentina, 125–26, 145–48, 151, 165, 197n52
Arkwright, Edwyn, 155–56
Arkwright, John Hungerford, 102–4, 155, 162
Artificial feed, 82, 86
Artificial selection, 29–30, 36
Australasia, 112–13, 119
Australasian Pastoralists' Review, 126, 129, 136
Australia: climate of, 51, 114; evolutionary isolation in, 9; first sheep in, 112; Hereford cattle in, 4, 21; meat from, 120, *121*, 122–23, 125–26; Merino sheep in, 17, 19, 77, 113–14; pastoral economy in, 6; refrigeration and, 120; surplus sheep in, 116–17; wool from, *61*, 113–16

Bailey, Joseph Russell, 104, 162–63, 212n34
Bakewell, Robert, 16, 25, 41–44, 55, 93, 130, 164, 192n112
Banks, Joseph, 60, 72–73
Bartley, Nehemiah, 60
Beef: American, 148, 150; Britishness and, 82, 140; demand for, 117–18, 197n69; fattening and, 81–87; feed and, 82, 86; in flow of commodities, 14–15; fraud with, 125–26, 200n131; frozen, 121–29; of Herefords, 81–88, 154–55; imports of, 118, 124–25; mutton vs., 118; price of, 117, 150; of Shorthorns, 83, 85, 88. *See also* Meat; Mutton
Belgian Blue cattle, 166

227

Belich, James, 179n48
Blackfaced, 25, 43, *44*
Books, herd, 103–5
Boreray sheep, 177n3
Boys, F., 49
Bradford, William, 10
Breeders, experience of, 35–37
Breeding: crossbreeding, 45, 65, 69, 99–100, 132–34, 138–39; experience and, 35–37; inbreeding, 28, 43, 88, 99–100, 136; selective, 5, 28, 35, 37, 74, 154, 159
Breeds: breeders and, 34–38; change in, 27–28; climate and, 31, 33–34, 66; "closed," 28; emergence of term, 28; focus on, 10–11; meat and, 46–47; "original" cattle, 93–95, *94*; pedigrees and, 88–89; in purebreeding, 28; regions and, 30–31; rise and fall of British, 4–8; variability and, 28–29. *See also* Native breeds
Breeds of the Domestic Animals of the British Islands (Low), 32, 42, 50
British Isles, 26
Britishness, 47–48, 50, 82, 118–19, 140
British Sheep Farming (Brown), 25, 26
Brooking, Tom, 114
Brown, William, 25, 26, 27, 33–34, 43, 48–49
Bruce, Alexander, 122–23
Bucke, Thomas George, 61, 69–70, 76

Cambridgeshire cattle, 30, 49–51
Campbell, J. H., 89, 95, 191n96
Canada, 21, 140, 146–47, 150–53, 163, 165, 168–70, 212n34
Caribbean, 38–39
Cassidy, Rebecca, 204n18
Castration, 159
Cattle, "original," 93–95, *94. See also specific breeds*
Cattle: Their Breeds, Management, and Diseases (Youatt), 82, 190n61
"Chain of breeding," 132
Chambers's Journal, 47, 115, 118, 123, 148

Change, in breeds, 27–28
Chaplin, Joyce, 13, 40
Charolais cattle, 166, 211n26
Cheviot sheep, 4; in Brown, 25; in Little, 35
Circulation, of commodities, 14–15
Climate: agricultural improvement and, 29, 48–49; in Australia, 77, 112–14; breed and, 31, 33, 66; of British Isles, 30–31, 33, 41, 69–70; of Canada, 152–53; crossbreeding and, 138; heredity vs., 27, 39, 49, 54–56, 66–67; Herefords and, 144, 152, 156; human evolution and, 39–40; imperialism and, 15; merinos and, 65–67; in Morocco, 155; mutton and, 123; nativeness and, 33–34, 78; of New Zealand, 129, 110–11, 114–15; transposition and, 39, 41; type and, 172; wool and, 77. *See also* Acclimatization
"Closed" breeds, 28
Clutton-Brock, Juliet, 83, 112
Coates, George, 98
Coats, of cattle, 92, 95–96
Coke, Thomas William, 76–77
Cold stores, 123–24. *See also* Refrigeration
Colley, Linda, 15
Collinson, Peter, 56, 65–66
Colonialism, 39–40, 111–16, 165–66
Color, of Herefords, 96–98, 154, 193n140
Commercial and Agricultural Magazine, 52, 87
Commodities, flow of, 14–15
Complete Grazier, The (Youatt), 85–86
Consumer preferences, meat and, 134–35
Consumption, 13–14
Continental System, 62
Cook, Les, 211n29
Cornish sheep, 45
Corn Laws, 117
Corriedale, The: New Zealand's Contribution to the Sheep World (Holford), 109

228 *Index*

Corriedale, The: New Zealand's Own Breed, 109
Corriedale-cross sheep, *137*
Corriedale sheep, 20, 109, *110*, 110–12, 138–39, 173, 202n196
Corriedale Sheep Society, 109
Cotswold sheep, 41, 166
County monikers, 90
Coventry, Andrew, 38
Craigie, P. G., 124
Creolization, 13
Cronon, William, 10, 189n27, 206n49
Crosby, Alfred, 8–9, 15
Crossbreeding, 45, 65, 69, 99–100, 132–34, 138–39
Culley, George, 78–79, 89–90, 92–93, 99, 191n96
Cultivator Middlesexiensis, 66, 76
Cumber, John, 211n23

Dartmoor, 41, 74
Darwin, Charles, 25, 28, 36, 99, 194n182
Darwin, Erasmus, 36
Davidson, William Soltau, 116, 135–36, 202n196
Derry, Margaret, 10, 36, 181n52, 193n149
Devon cattle, 7, 30, 79, 93–96, 140, 155, 157, 166, 170, 193n137
Devonshire cattle, 30, 89, 93, 95
Diet, meat in, 45–46. *See also* Meat; Milk
Dietetics, climate and, 40
Dishley sheep, *42*, 42–43. *See also* New Leicester
Dorsets, 41, 74
Dorsetshire sheep, 31, 114–15
Down sheep, in Brown, 25, 33
Duckham, Thomas, 79, 81, 85, 96–97, 145, 160, 204n18
Duncumb, John, 83, 193n137
Dutch Holsteins, 10

Ecological Imperialism (Crosby), 8
Ellman, T., 49

Empire, 8–15
Environment, heredity vs., 31–33. *See also* Acclimatization; Climate
Epigenetics, 180n22
Evolution, 25
Exmoor, 41, *68*
Experience, of breeders, 35–37
Exports, of cattle, 158–61
Eyton, Thomas Campbell, 95, 103, 160, 194n182, 194n185

Farmer's Magazine, 100–101, 151–52, 156, 159
Fattening, 81–87
Feed, artificial, 82, 86
Flanders cattle, 193n141
Fowler, John Kersley, 82, 96
France, 54, 57, 61–62
Franklin, Sarah, 112
Fraud, 125–29, 200n131
Free trade, 117
Frozen meat, 120–29, *127*

Garrard, George, 79
General Shorthorned Herd-Book (Coates), 98
General View of the Agriculture of the County of Cornwall (Worgan), 30
General View of the Agriculture of the County of Hereford (Duncumb), 83
Genotype, 29
George III, King, 17, 58–60
Gilliland, Samuel, 143
Gloucester, 41
Gooch, William, 30, 49–50
Grampians, 33
Grazing: breeding and, 29; imperialism and, 8
Grundy, Joan, 206n67, 207n90, 210n76

Hall, Catherine, 14
Hall, George Webb, 71
Hall, Steven J.G., 83
Hampton Court Herefords, 102
Harrison, Mark, 13

Index 229

Herd Book of Hereford Cattle, 89, 103–5, 160–61, 209n159
Herd books, 103–5, 211n13
Herdwick sheep, 31
Heredity, environment vs., 31–33
Hereford cattle, 4, 80, 84, 158, 169; Arkwright and, 155–56; artificial selection and, 153–54; in Australia, 4, 21; as British, 18–19; in Canada, 152–53, 168–70, 212n34; coloring of, 96–98, 154, 193n140; demand for, 159; in diversity of British livestock, 31; Hampton Court herd of, 102; Herefordshire and, 78; increasing recognition of, 143–44; Knight and, 95; as lacking prestige, 142–43; meat of, 81–88, 154–55; as native, 79–80, 170–71; pedigree of, 94–95, 103–4, 162–63, 174–75; purity of, 81, 88–95, 168–71, 174; Rare Breeds Survival Trust and, 168; Shorthorns vs., 18, 80–81, 98, 140, 143–44, 153, 155–56; success of, 172–73; Traditional, 21–22, 169, 170–75, 211n29; as transposable, 144–45; in United States, 140–41, 157–62, 207n85; "Volume 13 rule" and, 163–64
Hereford Cattle Society, 170–71
Hereford Herd Book Society, 102, 104, 160, 162–63
Herefordshire, 78
Higgins, David M., 200n131
Highlanders, 64
Hindson, J. C., 2, 174–75
History of Plymouth Plantation (Bradford), 10
Holford, G. H., 109, 139
Holmes, Matthew, 116
Horns, of cattle, 91, 93
Hoskyns, Hungerford, 102
Hunt, Charles Henry, 72, 74
Hunt, John, 36, 46, 53, 63–64, 71–73

Imperialism, 8–15, 111, 174, 212n41
Imports: of meat, 118, 124–25, 196n29; of wool, 60–61

Improvement. *See* Agricultural improvement
Inbreeding, 28, 43, 88, 99–100, 136
Intercolonial Stock Conference, 133
Ireland, 143, 146
Irish Farmers' Gazette, 102

Johnson, Frederik, 180n16
Jonsson, Fredrik Albritton, 64
Jordan, Terry G., 153
Journal of the Royal Agricultural Society of England, 124

King, Peter, 166–67, 172, 211n26
King George III, 17
Knight, Thomas Andrew, 95, 192n129, 193n141
Kyloe, 31, 85

Landrace sheep, 166
Lawrence, John, 28, 39–41
Leicester sheep, 41, 114, 130, 137, 166
Lillingston, Leonard, 125
Limousin cattle, 166
Lincoln Red cattle, 170
Lincoln sheep, 4, 114, 130, 134, 136–37
Little, James, 135, 202n196
Little, John, 35
Livestock Journal and Fancier's Gazette, 34, 91, 100–103, 105, 140–41, 143, 146–48, 152, 159–60
Longhorn cattle, 92, 93, 170, 192n112
Louis XVI, 57
Low, David, 32, 42, 50

Macarthur, John, 59
MacDonald, James, 149
Maori Wars, 113
Markings, of Herefords, 96–97, 154, 193n140
Marshall, William, 78, 93, 192n123
McArthur, John, 113
Meat, 45–51; refrigeration and, 119–22, 121, 198n84, 199n95; scarcity of, 117–19, 145–46. *See also* Beef; Mutton

230   *Index*

Medicine, climate in, 40
Melville, Elinor, 8–9, 178n20, 196n21
Mendelian inheritance, 181n44
Merino sheep, 53, 57, 128; absorption of, into British stock, 77; in Australia, 17, 19, 77, 113–14; Banks and, 52; climate and, 65–67, 69–70; crossbreeding of, 65, 69, 132–33, 136–37; debate over, 52–54; as diplomatic capital, 56–57; as extremophiles, 17; fattening of, 75–76; in France, 62; frozen meat trade and, 128–29; George III's flock of, 58; in Great Britain, 52–54, 58, 63–67, 69–70; "improvement" of, 74–75; mutton of, 52, 54–58, 71–76, 128; Negretti flock of, 58; New Leicester Longwool vs., 52–53; in New Zealand, 110, 114; prices of, 58–59; in show pen, 76–77; in Spain, 55–57, 62–63; in Sweden, 57; as "unimproved," 72–74; in United States, 196n31; upper class and, 60; war and, 54; wool of, 52, 54–58, 69–71, 111
Merino Society, 61–63, 69
Middle class, 45
Milk, 85–86, 101. *See also* Meat
Milk: A Local and Global History (Valenze), 10
Milk Marketing Board, 163–64, 211n26
Miller, T. L., 161–62, 208n130
Mintz, Sydney, 14
Morocco, 155
Mouflon, 1
Mutton: Australian, 121, 123, 125, 128–29; beef vs., 118; Britishness and, 118–19; consumer preferences and, 134–35; of Corriedales, 109, 111; demand for, 81–82, 117–19, 197n69; in flow of commodities, 14–15; fraud with, 125–26, 200n131; frozen, 121–29, 127; imports of, 118, 121, 124–25, 196n29; of Merinos, 52, 54–58, 71–77, 128; of New Leicester Longwool, 63–64; price of, 117; shift to, from wool, in New Zealand, 129–30; wool vs., 55–58. *See also* Beef; Meat

Naming, 90
Napoleon, 62
Native breeds, 25; in Americas, 149–50, 153; Britishness and, 128; of cattle, 78–79; climate and, 67; conservation and, 168; crossbreeding and, 65, 74; Hereford as, 140, 143, 170–71, 193 (n. 141); "improvement" and, 25, 43–45; invention of, 3; meaning of, 15, 27; merino vs., 71, 74–77; in New Zealand, 111–12, 134, 138–39; population of, 25; Rare Breeds Survival Trust and, 166; Soay as, 1–3; wool and, 55
Nativeness: acclimatization and, 12–13; climate and, 33–34, 78; diversity and, 95; inheritance and, 33; meaning of, 11–13, 22, 27, 90–91, 173–74; purity and, 88–95, 97; of Shorthorns, 81
Naturalization, 70–71
Natural selection, 25, 36
Negretti, 58
New Leicester Longwool, 16, 25, 41–44, 42, 48–49, 52–55, 63–69, 72, 85
New Review, 124
New Zealand: climate of, 129, 110–11, 114–15; colonization of, sheep and, 112–17; population of sheep in, 115; sheep breed for, 133–39, 137; surplus sheep in, 116–17; topography of, 183n120
New Zealand and Australia Land Company, 116, 120, 136, 202n196
New Zealand Country Journal, 115
New Zealand Farmer, 36, 109, 111, 115, 118, 129–31, 134–36
Norfolk cattle, 45, 74, 93
North Devons, 30, 41

Observations on Live Stock (Culley), 78–79, 92–93, 99
"Original" cattle, 93–95, 94

Origin of Species, The (Darwin), 25
Ovis aries, 1, 9, 112
Ovis musimon, 1

Parkinson, Richard, 190n60
Parry, Caleb Hilliar, 53, 60, 73–74
Pasture, 29
Paular flock, 63
Pawson, Eric, 114
Peak District, 34
Peden, Robert, 10
Pedigrees, 88–89, 98, 100–105, 160–61
Phenotype, 29, 96–98
Population growth, 45
Portugal, 58
Powell, Percy, 104
Practical Observations on the Improvement and Management of Mountain Sheep, and Sheep Farms (Little), 35
Practical Treatise on the Merino and Anglo-Merino Breeds of Sheep, A (Hunt), 72
Prairie Cattle Company of West Texas, 205n42
Price, of meat, 117, 150
"Progeny test," 181n52
Purebreeding, 28
Purity, 81, 88–95, 98–102

"Q" (writer), 90

Race, 40–41
Rare Breeds Survival Trust (RBST), 1–2, 166–68, 170, 172, 175
RBST. *See* Rare Breeds Survival Trust (RBST)
Records of Old Times: Historical, Social, Political, Sporing and Agricultural (Fowler), 82
Red Poll cattle, 211n26
Refrigeration, 119–23, *121*, 198n84, 199n95
Regions, breeds and, 30–31
Remarks on Live Stock and Relative Subjects (Coventry), 38

Reynall, R. W., 143
Ritvo, Harriet, 10, 154, 192n118
Roberts, John, 133–34
Romney Marsh sheep, 31, 114
Romney sheep, 4, 20, 136
Royal sheep, 58–60
Rudge, Thomas, 47
Rural Economy of Gloucestershire (Marshall), 78
"Rusticus" (writer), 132
Rutabaga, 29
Ryeland sheep, 4, 31, 41, 55, 66, 74, 75, 78, 166

St. Kilda, 1–2, 177n3–177n4
Sayre, Nathan, 10
Scotland: climate in, 34; graving and, 29
Scots Magazine, 39
Scottish Enlightenment, 64
Scottish Highland, 85
Seasoning, 40
Sebright, John, 28, 37–38, 65, 67
Selective breeding, 5, 28, 35, 37, 74, 154, 159
Semen, 164, 211n26
Seven Years' War, 64
Shapin, Steven, 15
Shetlands, 31
Shorthorn cattle, 4, *91*, 142, 170, 204n18; degeneration of, 145; herd book for, 211n13; Herefords vs., 18, 80–81, 98, 140, 143–44, 153, 155–56; "improvement" of, 167, 192n112; meat of, 83, 85, 88; pedigrees of, 100–101, 193n149; popularity of, 101–2, 141–43; population of, 212n31; purity of, 90, 95–96, 98, 163; in United States, 151–53, 207n85
Shropshire sheep, 31, 129–30, 137
Sinclair, John, 29, 33, 43, 46
Six Weeks' Tour through the Southern Counties of England and Wales, A (Young), 180n16
Smithfield Club, 87, 189n35
Soay sheep, 1–3, 177n3

232   *Index*

Somersetshire sheep, 41
Somerville, John Southey, 29, 60–61, 72
Sotham, W. H., 149, 152
South Africa, 15–16, 166, 209n140
South America, 146, 159, 163
South Devons, 30, 41
Southdown sheep, 31, 41, 49, 50, 129–30, 134, 137, 182n83
Spain, 55–58, 62–63, 165. *See also* Merino sheep
Spanish wool, 64–71
Steel, Frances, 179n48
Stokes, Evelyn, 174
Surplus sheep, 116–17
Sussex cattle, 93
Swede, 29
Sweden, 57

Tallow, 117
Temperament, 84–85
"Texas type," 149–50
Texel sheep, 166
Thatcher, Thomas, 136
Thompson, Benjamin, 60, 62–64, 69, 76
Tollet, George, 60
Topography, 15–22
Traits: consistent replication of, 88–89; discerning of, 37–38
Transhumantes, 56, 73
Transmutation, 38–41
Transposition, 38–41, 173; Empire and, 38–39; human, 39–40; nativeness and, 2; type and, 141–45
Treaty of Waitangi, 113
Trueness, 95–98
Tudge, William, 204n18
Turner, G. T., 100
Type. *See* Breeds; Native breeds

United States, 21, 119, 140–41, 146–53, 157–62, 165, 205n40, 205n45, 205n48, 207n90; Merinos in, 196n31
Urwick, S. W., 162

Valenze, Deborah, 10
Valle de Mezquital, 178n20
Variability, 28–29
Variation of Animals and Plants under Domestication (Darwin), 28, 36
Vermont, 196n31
Vernon, James, 183n94
Vialles, Noelie, 14
"Volume 13 rule," 163–64

Warehouses, 123
Welsh sheep, 74
Westcar, Joseph, 82, 95, 189n35, 190n79
West Highlander, 79
Weston, T., 86–88, 190n79
White, Taylor, 134, 137–38, 203n201
White Park cattle, 170
"Wild" cattle, 93–95, 94
Wiltshire sheep, 31, 41, 74
Wool: Australian, 61, 113–16; breeding and, 64–65; climate and, 33–34; colonialism and, 116; Corriedale, 111; crossbreeding and, 132; demand for, 60–61, 61; imports of, 60–61; meat vs., 55–58; Merino, 52, 54–58, 60–61, 69–71, 111, 114, 117; shift from, to meat, in New Zealand, 129–30; Spanish, 64–71
Worcester, Don, 205n42, 207n90
Worgan, George B., 30–31, 45

Youatt, William, 79–80, 82, 85–86, 90–95, 190n61
Young, Arthur, 29, 52, 89, 180n16